Grade 6 · Unit 1

Inspire Science

Life Structure and Function

Phenomenon: Why can you see through these fish?

The Glass Catfish, or *Kryptopterus vitreolus*, is a nearly transparent species native to the waters of Thailand. Some internal body systems, such as digestive, skeletal, and nervous, are easily identifiable.

If you look at a Glass Catfish with a magnifying lens, you can see its heart beating!

FRONT COVER: (t)Grigorev Mikhail/Shutterstock, (bl)279photo/iStock/Getty Images, (br) STEVE GSCHMEISSNER/SCIENCE PHOTO LIBRARY/Getty Images. **BACK COVER:** Grigorev Mikhail/Shutterstock.

mheducation.com/prek-12

Send all inquiries to:
McGraw-Hill Education
STEM Learning Solutions Center
8787 Orion Place
Columbus, OH 43240

ISBN: 978-0-07-687330-2
MHID: 0-07-687330-7

Printed in the United States of America.

4 5 6 7 8 9 QSX 23 22 21 20 19

McGraw-Hill is committed to providing instructional materials in Science, Technology, Engineering, and Mathematics (STEM) that give all students a solid foundation, one that prepares them for college and careers in the 21st century.

Welcome to

Inspire Science

Explore Our Phenomenal World

Learning begins with curiosity. Inspire Science is designed to spark your interest and empower you to ask more questions, think more critically, and generate innovative ideas.

Start exploring now!

Inspire Curiosity • Inspire Investigation • Inspire Innovation

Authors, Contributors, and Partners

Program Authors

Alton L. Biggs
Biggs Educational Consulting
Commerce, TX

Ralph M. Feather, Jr., PhD
Professor of Educational Studies and Secondary Education
Bloomsburg University
Bloomsburg, PA

Douglas Fisher, PhD
Professor of Teacher Education
San Diego State University
San Diego, CA

Page Keeley, MEd
Author, Consultant, Inventor of Page Keeley Science Probes
Maine Mathematics and Science Alliance
Augusta, ME

Michael Manga, PhD
Professor
University of California, Berkeley
Berkeley, CA

Edward P. Ortleb
Science/Safety Consultant
St. Louis, MO

Dinah Zike, MEd
Author, Consultant, Inventor of Foldables®
Dinah Zike Academy, Dinah-Might Adventures, LP
San Antonio, TX

Advisors

Phil Lafontaine
NGSS Education Consultant
Folsom, CA

Donna Markey
NBCT, Vista Unified School District
Vista, CA

Julie Olson
NGSS Consultant
Mitchell Senior High/Second Chance High School
Mitchell, SD

Content Consultants

Chris Anderson
STEM Coach and Engineering Consultant
Cinnaminson, NJ

Emily Miller
EL Consultant
Madison, WI

Key Partners

American Museum of Natural History
The American Museum of Natural History is one of the world's preeminent scientific and cultural institutions. Founded in 1869, the Museum has advanced its global mission to discover, interpret, and disseminate information about human cultures, the natural world, and the universe through a wide-ranging program of scientific research, education, and exhibition.

PhET Interactive Simulations
The PhET Interactive Simulations project at the University of Colorado Boulder provides teachers and students with interactive science and math simulations. Based on extensive education research, PhET simulations engage students through an intuitive, game-like environment where students learn through exploration and discovery.

SpongeLab Interactives
SpongeLab Interactives is a learning technology company that inspires learning and engagement by creating gamified environments that encourage students to interact with digital learning experiences. Students participate in inquiry activities and problem-solving to explore a variety of topics through the use of games, interactives, and video while teachers take advantage of formative, summative, or performance-based assessment information that is gathered through the learning management system.

Measured Progress, a not-for-profit organization, is a pioneer in authentic, standards-based assessments. Included with New York Inspire Science is **Measured Progress STEM Gauge®** assessment content which enables teacher to monitor progress toward learning NGSS.

Table of Contents
Life Structure and Function

Module 1 Cells and Life

Module 2 Body Systems

Cells and Life

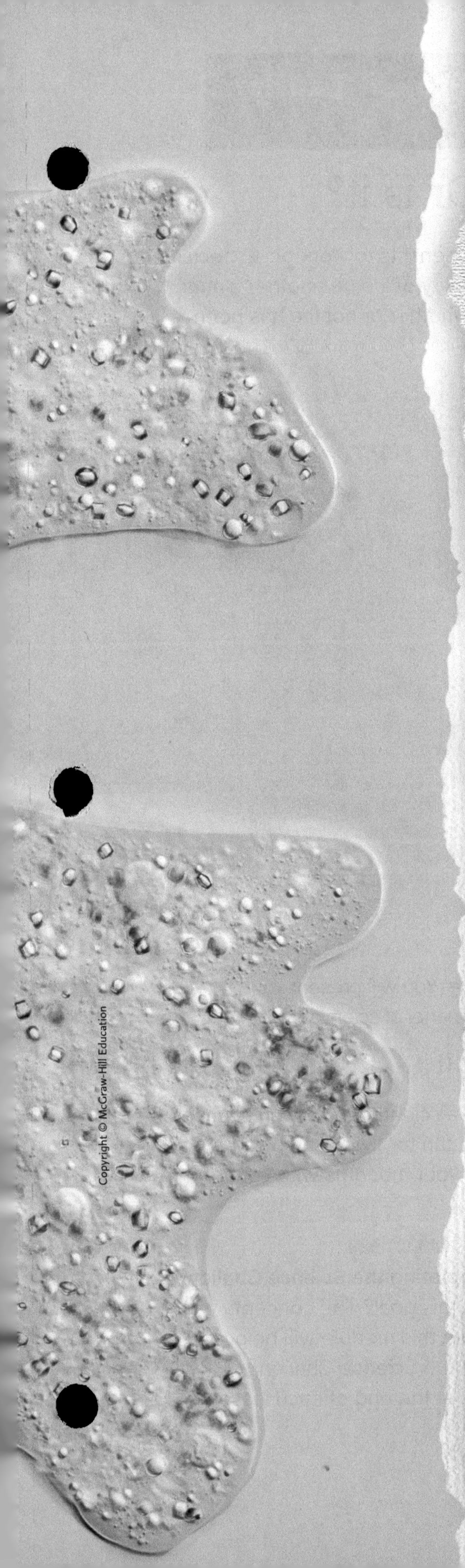

ENCOUNTER THE PHENOMENON

How does this microscopic amoeba perform all the same functions that you do to stay alive?

GO ONLINE
Watch the video *What's for Lunch?* to see this phenomenon in action.

Collaborate With a partner, discuss the functions that both you and the amoeba in the photo need to perform in order to stay alive. How do you think those functions are carried out? Record or illustrate your thoughts in the space below.

STEM Module Project Launch
Science Challenge

IT'S ALIVE! Or is it?

You have been invited by scientists to work on a special project involving the possibility of life on another planet. Your job is to help decide whether or not life has been found in outer space.

Astronauts were able to travel to a nearby planet and gather samples. Some of the samples appear to be living, but the scientists don't know for sure.

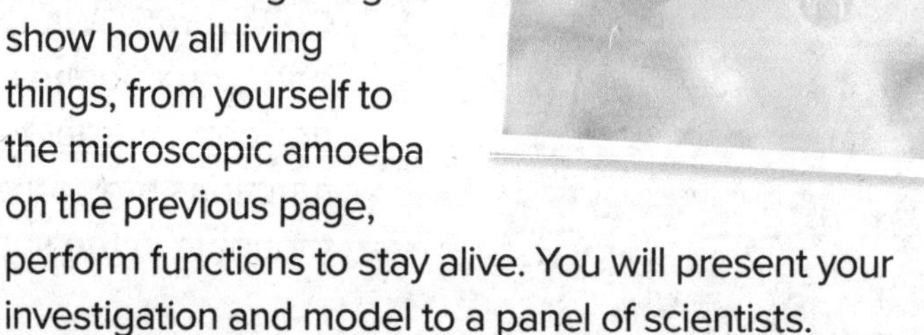

Your goal is to determine whether or not the samples are living by conducting an investigation to show what living things are made of, and developing and using a model of the building blocks of a living thing to show how all living things, from yourself to the microscopic amoeba on the previous page, perform functions to stay alive. You will present your investigation and model to a panel of scientists.

Start Thinking About It

In the photo above you see a scientist observing samples. How do you think scientists can tell living samples from nonliving samples? Discuss your thoughts with your group.

STEM Module Project

Planning and Completing the Science Challenge
How will you meet this goal? The concepts you will learn throughout this module will help you plan and complete the Science Challenge. Just follow the prompts at the end of each lesson!

SCIENCE PROBES

Are seeds alive?

Four friends were planting flowers in the school garden. They began to question whether or not seeds are alive. Here are their thoughts:

Eli: I don't think seeds are alive until they are watered.

Tory: I think seeds are always alive.

Kelly: I don't think seeds are alive.

DeAndre: I don't think seeds are alive until they sprout.

Circle the friend you most agree with. Explain why you agree with that person.

You will revisit your response to the Science Probe at the end of the lesson.

LESSON 1

Exploring Life

ENCOUNTER THE PHENOMENON | How can you tell whether or not this campfire is alive?

GO ONLINE

Watch the video *Dancing Flames* to see this phenomenon in action.

All living things share the same basic characteristics that make them different from nonliving things. Brainstorm what characteristics all living things might have. Record your thoughts below.

Observe the lit candles. Record your observations below. Are there any of the characteristics that you listed above that you also observe in the flames?

What characteristics do the flames share with living things? What characteristics of living things do the flames lack? Record your thoughts below.

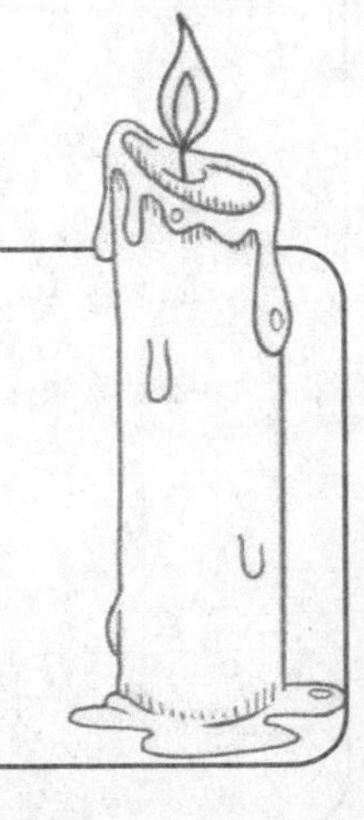

EXPLAIN THE PHENOMENON

You observed flames and compared them to living things. Did you notice that all living things must have some characteristics in common that make them different from nonliving things? Now make a claim about characteristics that all living things share.

CLAIM

All living things...

COLLECT EVIDENCE as you work through the lesson. Then return to these pages to record your evidence.

EVIDENCE

A. What evidence have you discovered to explain what living things are made of that differentiates them from nonliving things, such as a flame?

MORE EVIDENCE

B. What evidence have you discovered to explain the characteristics of life that differentiate living things from nonliving things, such as a flame?

When you are finished with the lesson, review your evidence. If necessary, based on the evidence, revise your claim.

REVISED CLAIM

All living things...

Finally, explain your reasoning for how and why your evidence supports your claim.

REASONING

The evidence I collected supports my claim because...

What are living things?

How can you tell if something is living or nonliving? What is the difference between living things and nonliving things?

INVESTIGATION

Living v. Nonliving

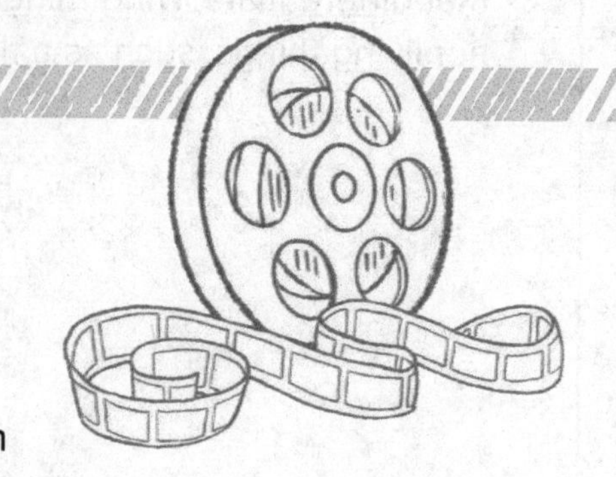

GO ONLINE Watch the video *Is it alive?*

Use the table below to decide which things are living and which are not. Explain your reasoning for each.

	Living	Nonliving	Reasoning
Kit fox			
Volcano			
Grapes			
Sand/water			
Sponge			
Robot			
Chrysalis			
Eggs			
Moon/clouds			
Sea anemone			
Light rays			
Bacteria			

Want more information?
Go online to read more about the characteristics of living things.

FOLDABLES
Go to the Foldables® library to make a Foldable® that will help you take notes while reading this lesson.

Building Blocks of Life All living things share seven characteristics of life. The first characteristic that living things have in common is what they are made of. Let's take a closer look at the building blocks of life.

LAB A Closer Look at Life

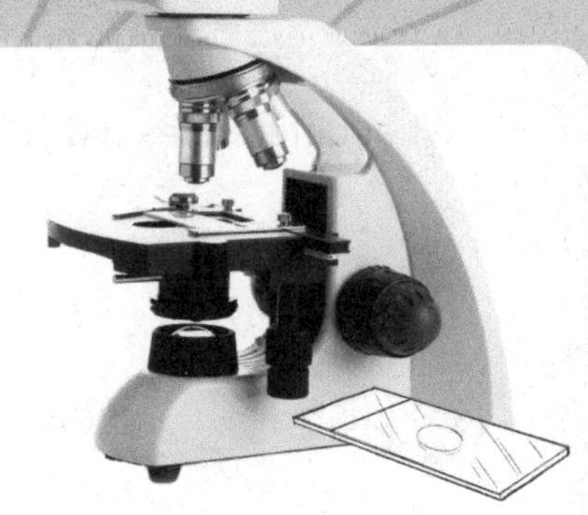

Safety

Materials

microscope

prepared slides of human cheek sample, onion, pond water, salt

Procedure

1. Read and complete a lab safety form.
2. Observe each slide under the microscope. What do you see? Illustrate and record your observations for each slide in the Data and Observations section below.
3. Follow your teacher's instructions for proper cleanup.

Data and Observations

Cheek sample	*Onion*
Pond water	*Salt*

Analyze and Conclude

4. What similarities did you notice between the slides?

5. What differences did you notice between the slides?

6. Make a claim about the differences between the living samples and the nonliving samples.

7. What evidence from the investigation supports your claim?

Technology Leads to Discovery You just used a microscope to observe what living things are made of—**cells.** Have you ever looked up at the night sky and tried to find other planets in our solar system? It is hard to see them without using a telescope, because other planets are millions of kilometers away. Just like we can use telescopes to see other planets, we can use microscopes to see cells. But people didn't always know about cells. Because cells are so small, early scientists had no tools to study them. It took hundreds of years for scientists to learn about cells.

Have you ever used a magnifying lens to see details of an object? If so, then you have used a tool similar to the first microscope. The invention of microscopes enabled people to see details of living things that they could not see with the unaided eye. The microscope was an advance in engineering that enabled people to make important discoveries about living things.

For centuries, people have been looking for ways to see objects in greater detail. Can you find a way to do this using only the simple materials in the lab below?

Magnify It

Safety    

Materials

newspaper
water
dropper
glass jar
plastic 2-liter bottle
scissors
plastic wrap

Procedure

1. Read and complete a lab safety form.
2. Before you begin exploring the materials, define the problem.

3. What are the criteria and constraints for your exploration? Record them below.

Criteria:	Constraints: Time: Materials:

Procedure, continued

4. With your group, brainstorm how you could use the given materials to magnify objects. Record your ideas below.

5. Test your ideas and collect data. What worked and didn't work? Record your observations in the Data and Observations section.

6. Follow your teacher's instructions for proper cleanup.

Data and Observations

Analyze and Conclude

7. Evaluate each of your design solutions. Looking at your data and thinking about what worked best, record your best solution below. Or, combine aspects of different solutions to propose your best solution.

Read a Scientific Text

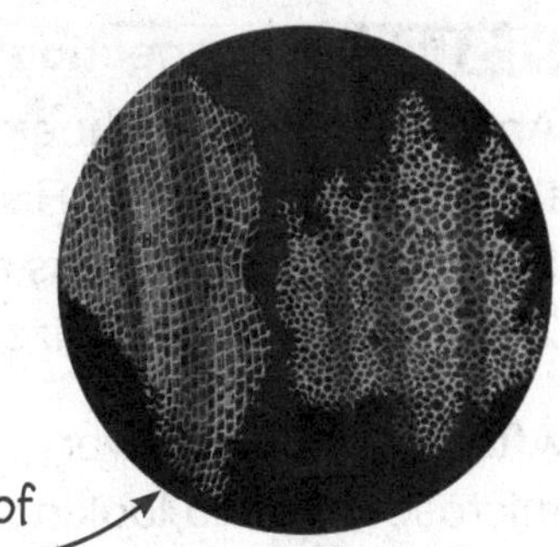

HISTORY Connection More than 300 years ago, an English scientist named Robert Hooke built a microscope. He used that microscope to discover cells. How did he do it, and what conclusions did he come to? Read his original writings on observations of his discovery below.

Take a look at this colorized photo of one of Robert Hooke's drawings!

CLOSE READING

Inspect

Read the passage *Observation XVIII of the Texture of Cork.*

Find Evidence

Reread the passage. Underline words and phrases in which Hooke describes the cells of the cork.

Make Connections

Talk About It With your partner, discuss how the microscope led to the discovery of cells.

PRIMARY SOURCE

Observation XVIII of the Texture of Cork
Robert Hooke, 1665

I took a good clear piece of Cork, and with a Pen-knife sharpen'd as keen as a Razor, I cut a piece of it off, and thereby left the surface of it exceeding smooth, then examining it very diligently with a Microscope, me thought I could perceive it to appear a little porous; but I could not so plainly distinguish them, as to be sure that they were pores, much less what Figure they were of: But judging from the lightness and yielding quality of the Cork, that certainly the texture could not be so curious, but that possibly, if I could use some further diligence, I might find it to be discernable with a Microscope, I with the same sharp Penknife, cut off from the former smooth surface an exceeding thin piece of it, and placing it on a black object Plate, because it was itself a white body, and casting the light on it with a deep plano-convex Glass, I could exceeding plainly perceive it to be all perforated and porous, much like a Honey-comb, but that the pores of it were not regular; yet it was not unlike a Honey-comb in these particulars.

First, in that it had a very little solid substance, in comparison of the empty cavity that was contain'd between [...]

Next, in that these pores, or cells, were not very deep, but consisted of a great many little Boxes [...]

Source: Project Gutenberg

ENGINEERING Connection How have microscopes helped people learn about living things on a different scale?

HISTORY Connection In the late 1600s, the Dutch merchant Anton van Leeuwenhoek (LAY vun hook) made improvements to the first microscopes. His microscope, similar to the one shown in the image, had one lens and could magnify an image about 270 times its original size. This made it easier to view organisms.

Anton van Leeuwenhoek observed pond water and insects using a microscope like the one shown below.

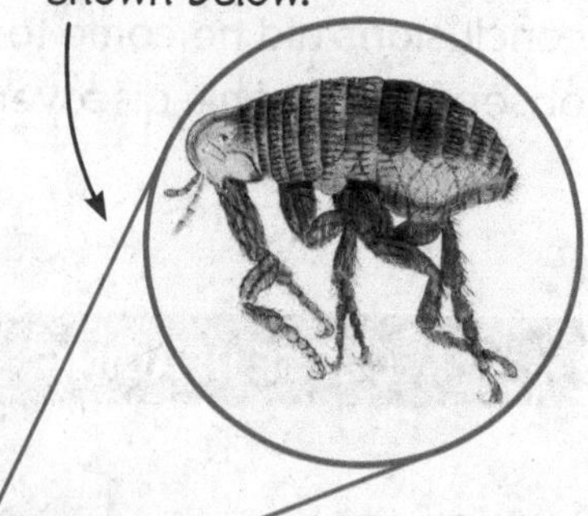

After Hooke's discovery, other scientists began making better microscopes and looking for cells in many other places, such as pond water and blood. The newer microscopes enabled scientists to see different structures inside cells. Three important observations about cells made by three different scientists were combined into one theory called the **cell theory.**

INVESTIGATION

Discovering the Cell Theory

GO ONLINE Watch the animation *The Cell Theory.*

As you watch the animation, fill in the table on the three principles of the cell theory below.

The Cell Theory	
Principle	**Example**

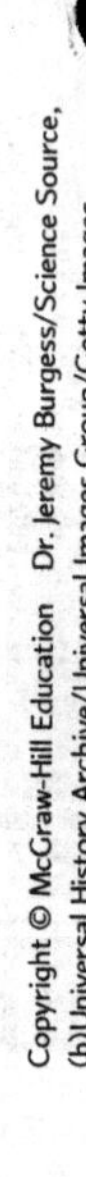

Principles of the Cell Theory You might recall that all matter is made of atoms and that atoms combine and form molecules. Molecules make up cells. All living things are made up of cells, which are the smallest unit of life. Cells perform different functions to keep organisms alive. All cells come from preexisting cells through the process of cell division.

COLLECT EVIDENCE

What are living things made of that differentiates them from nonliving things, such as a flame? Record your evidence (A) in the chart at the beginning of the lesson.

ENGINEERING Connection Since the development of the cell theory in the 1830s, microscopes have continued to become more advanced. If you have used a microscope in school, then you have probably used a light microscope. **Light microscopes** use light and lenses to enlarge an image of an object. Light microscopes can enlarge images up to 1,500 times their original size. In some cases, the object, such as the blood cells in the photo below, must be stained with a dye in order to see any details.

You might know that electrons are tiny particles inside atoms. **Electron microscopes** use a magnetic field to focus a beam of electrons through an object or onto an object's surface. An electron microscope can magnify an image up to 100,000 times or more. The two main types of electron microscopes are transmission electron microscopes (TEMs) and scanning electron microscopes (SEMs).

TEMs are usually used to study extremely small things such as cell structures. In a TEM, electrons pass through the object and a computer produces an image of the object. A TEM image of a white blood cell is shown below.

SEMs are usually used to study an object's surface. In an SEM, electrons bounce off the object and a computer produces a three-dimensional image of the object. An image of blood cells from an SEM is shown below. Note the difference in detail in this image compared to the image of blood cells from a light microscope.

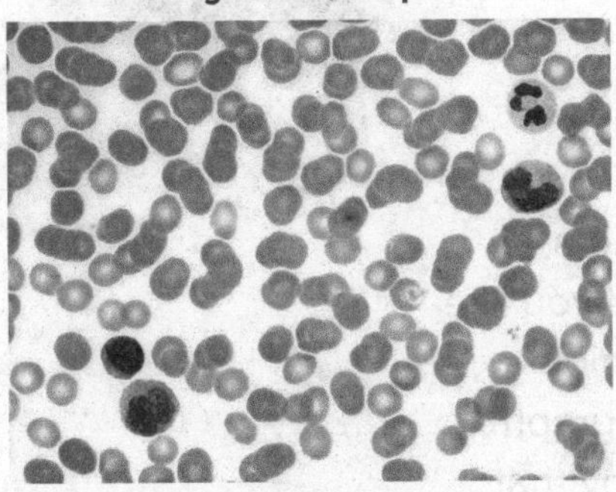

Stained LM Magnification: 640×

Transmission Electron Microscope

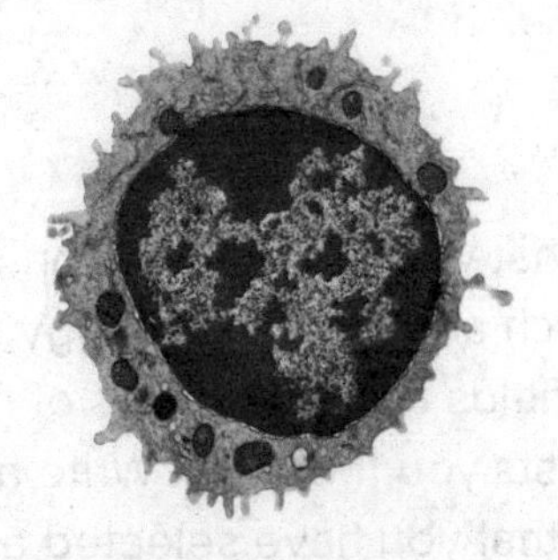

Color-Enhanced TEM Magnification: 8,900×

Scanning Electron Microscope

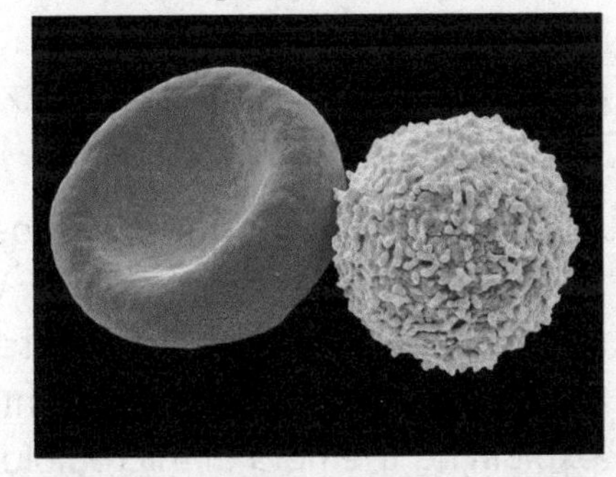

Color-Enhanced SEM Magnification: 8,500×

STEM Careers

A Day in the Life of a Microbiologist

Microbiologists study living things that are too small to be seen with the unaided eye, such as bacteria, algae, and fungi. Some microbiologists also study viruses. Without microscopes, the field of microbiology and the industries it supports, such as the food and medical industries, would not be the same today.

A typical day in the life of a microbiologist depends on what specific field of microbiology he or she works in. Some microbiologists focus solely on certain organisms, such as bacteria; some focus on the ways in which microorganisms interact with the environment; and some focus on ways to detect, treat, and prevent diseases caused by microorganisms.

Microbiologists spend much of their time preparing the samples that they study, conducting experiments, and writing reports on their findings. Since their specimens cannot be seen with the unaided eye, microbiologists use microscopes, along with many other technologies, in their work.

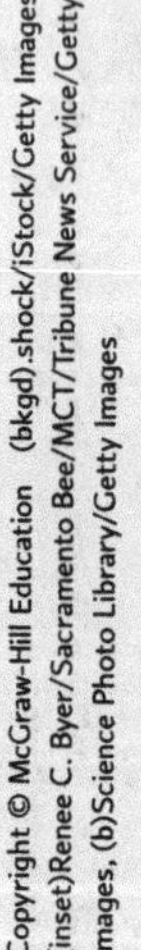

It's Your Turn

WRITING Connection Suppose that you are planning on getting a college degree in microbiology. Which area of microbiology would you like to focus on? Research the different fields and specialties of microbiology, and then choose the one that interests you the most. Write a paragraph explaining the field of microbiology that you have selected and why that field interests you.

How many cells do living things have?

Organisms are organized in different ways. Living things that are made of only one cell are called **unicellular organisms.** Living things that are made of two or more cells are called **multicellular organisms.**

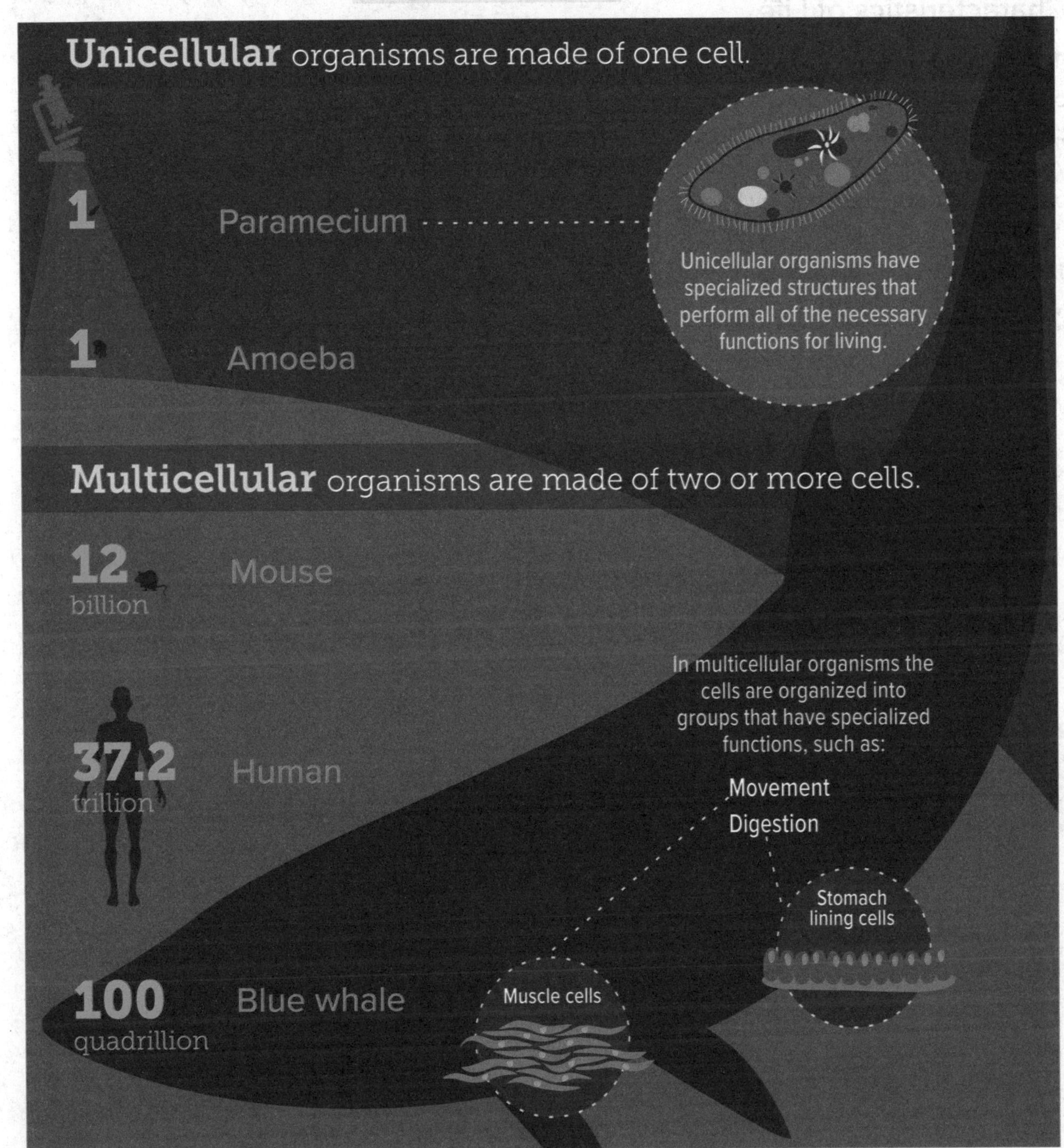

THREE-DIMENSIONAL THINKING

What do you notice about **scale, proportion, and quantity** of cells in unicellular versus multicellular organisms? Record your response in your Science Notebook.

Common Characteristics You now know that all living things are made of cells. What are the other six characteristics that all living things have in common? These include organization, growth and development, reproduction, response to stimuli, maintaining internal conditions, and using energy.

INVESTIGATION

Characteristics of Life

WRITING Connection Your teacher will assign one of the characteristics of life for you to research with your group. Use the graphic organizer for your characteristic to help guide your research. Your group will create a visual for your assigned characteristic and present it to the class. Fill in the graphic organizers for the rest of the characteristics as the other groups present.

1. Organization

Living things are organized by...

______ organisms are less complex.

______ organisms are more complex.

2. Growth and development

Growth is...	Development is...

3. Reproduction

Reproduction is...

Types of reproduction:

4. Response to stimuli

Internal stimuli are...

Examples:

External stimuli are...

Examples:

5. Maintaining internal conditions

Maintaining internal conditions is called ______________________.

Examples of how organisms maintain internal conditions:

6. Use of energy

Most organisms get their energy from ______________________.

______________ use energy from the Sun to make food.

Other organisms obtain energy by ______________________ ______________________.

Plan your visual and presentation below.

Characteristics of Life All living things are organized according to different structures that perform different functions. Living things grow and develop, meaning they increase in size and go through changes during their lifespans. Living things create new living things through the process of **reproduction.** They also respond to changes in their environments, called stimuli. Another characteristic of organisms is **homeostasis**, which is the ability to maintain steady internal conditions when outside conditions change. All organisms require energy for everything they do. If something doesn't display each of these characteristics, it is not a living thing.

COLLECT EVIDENCE

What are the characteristics of living things that differentiate them from nonliving things, such as a flame? Record your evidence (B) in the chart at the beginning of the lesson.

What are the different types of cells?

Recall that the use of microscopes enabled scientists to discover cells. With more advanced microscopes, scientists discovered that all cells can be grouped into two categories—prokaryotic (proh ka ree AH tihk) cells and eukaryotic (yew ker ee AH tihk) cells.

All cells contain genetic material—the means by which information is transmitted from one generation to the next. In some cells this genetic material is surrounded by a lining. In **prokaryotic** cells, the genetic material is not surrounded by a lining. This is the most important feature of a prokaryotic cell. In general, prokaryotic cells are smaller than eukaryotic cells and have fewer parts to their cells. Most prokaryotic cells are unicellular organisms and are called prokaryotes. Some prokaryotes live in small groups called colonies. Some can also live in extreme environments.

Plants, animals, fungi, and organisms called protists are all made of eukaryotic cells, and are called eukaryotes. With few exceptions, each **eukaryotic** cell has genetic material that is surrounded by a lining. Every eukaryotic cell also has other structures, called **organelles,** which have specialized functions. Most organelles are surrounded by linings. Eukaryotic cells are usually larger than prokaryotic cells. About ten prokaryotic cells would fit inside one eukaryotic cell.

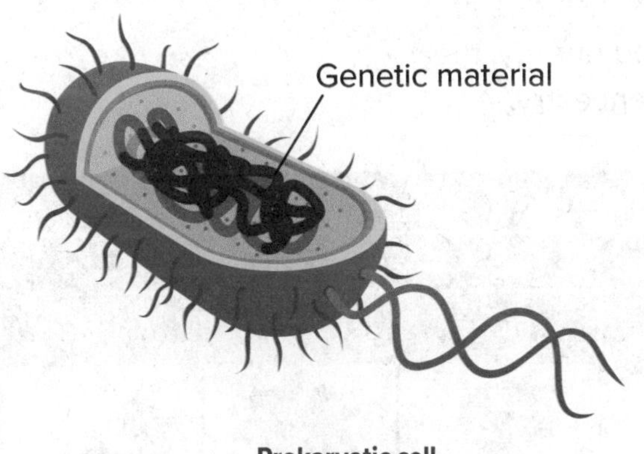

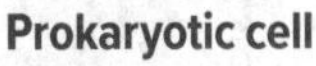
Prokaryotic cell

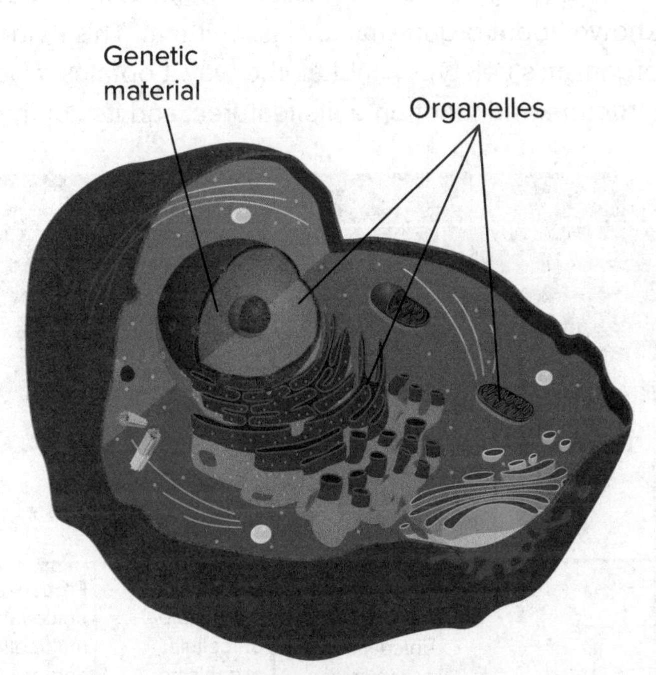

Eukaryotic cell

Compare and contrast prokaryotic cells and eukaryotic cells by completing the Venn diagram on the next page.

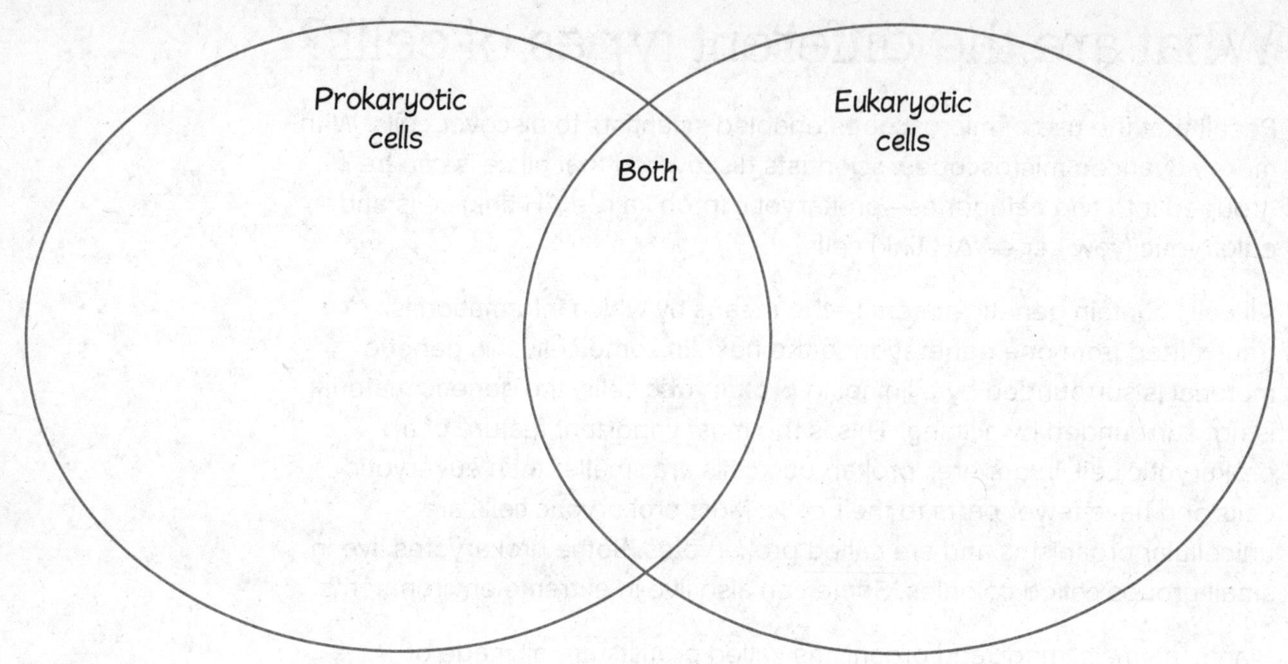

Classification Organisms are classified according to their cell type—prokaryotic or eukaryotic, as well as other characteristics. All organisms are classified into one of three domains—Bacteria, Archaea, or Eukarya (yew KER ee uh)—and then into one of six kingdoms. Organisms in the Bacteria and Archaea domains are unicellular prokaryotes, while organisms in the Eukarya domain are eukaryotes. The classification system of living things is still changing. The current classification method uses all the evidence that is known about organisms to classify them. This evidence includes an organism's cell type, habitat, the way it obtains food and energy, the structure and function of its features, and its common ancestry.

Domains and Kingdoms

Domain	Bacteria	Archaea	Eukarya			
Kingdom	**Bacteria**	**Archaea**	**Protista**	**Fungi**	**Plantae**	**Animalia**
Example						
Characteristics	Bacteria are simple unicellular organisms.	Archaea are simple unicellular organisms that often live in extreme environments.	Protists are unicellular or multicellular and are more complex than bacteria and archaea.	Fungi are unicellular or multicellular and absorb food.	Plants are multicellular and make their own food.	Animals are multicellular and take in their food.

What about viruses? Where do they fit into this classification system? Are they living things?

A Closer Look: Are viruses living things?

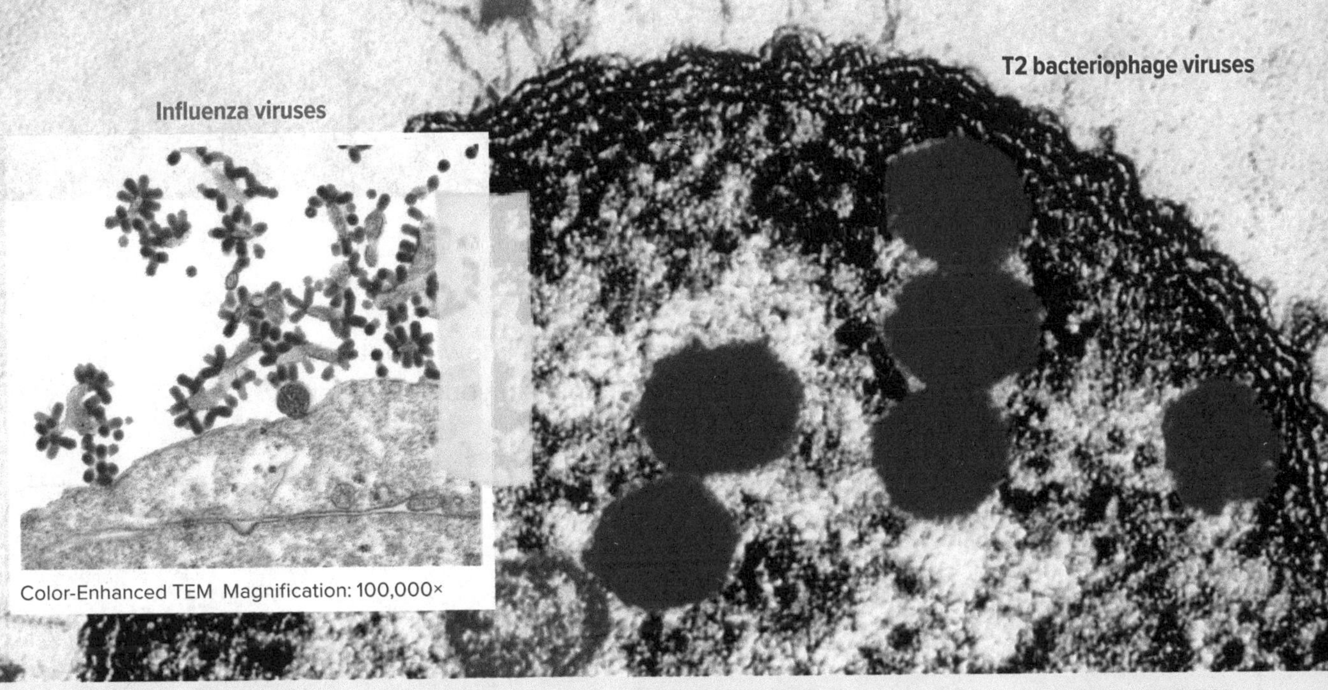

Color-Enhanced TEM Magnification: 100,000×

Color-Enhanced TEM Magnification: 64,000×

Have you ever had a cold, the chicken pox, or the flu? If so, then you have been infected by a virus. Are viruses alive? Let's see if viruses have all the characteristics of life to find out more.

Like living things, viruses also have structure and are organized. Viruses don't grow or develop like living things do. Viruses do reproduce and make more viruses, but in order to do so they need to enter and take over living cells. While viruses don't respond to light or sound the way that animals do, we still cannot say for sure that viruses don't respond to anything in their environment. It takes a lot of energy for viruses to replicate, but the energy that viruses use comes from the host. Viruses are not made up of cells like living things are, so they do not have the parts of cells required to maintain certain internal conditions.

Most scientists agree that since viruses aren't made out of cells, don't grow, use energy only from a host, and don't maintain homeostasis, they are not living things.

Ebola virus particles

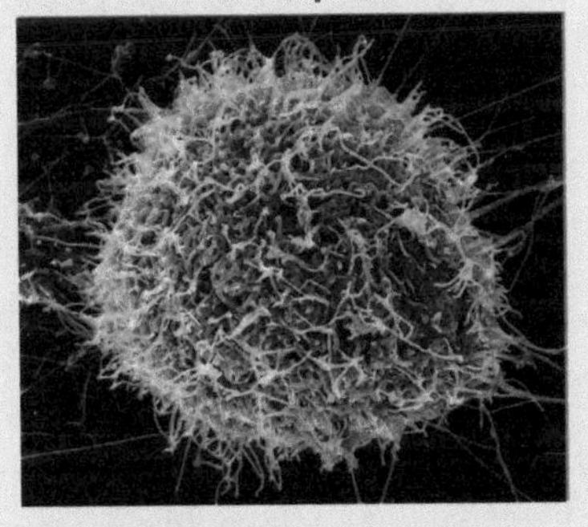
Color-Enhanced SEM Magnification: Unavailable

It's Your Turn

WRITING Connection What questions do you still have about viruses? Research one question, gathering information from several sources, and write a paragraph on your findings. Record other questions that come up as your conduct your research.

LESSON 1

Review

Summarize It!

1. **Organize** Create an infographic that shows what you know about living things. Include illustrations, key terms, and examples in your infographic. Make sure to show all that you know!

Three-Dimensional Thinking

2. If a living organism contains a cell with the structures seen below, which of the following can you conclude about the organism?

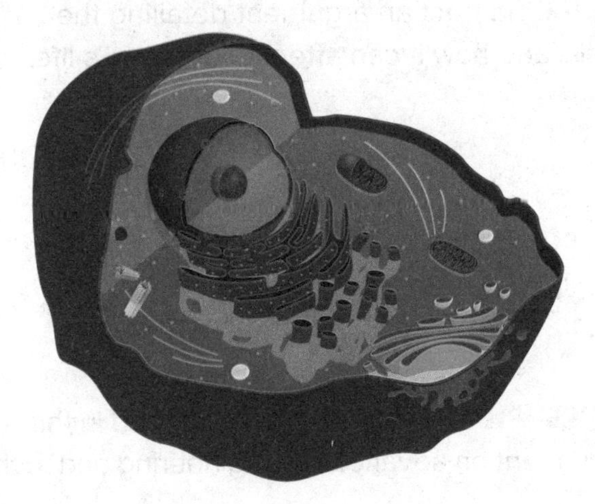

A The organism is a eukaryote.

B The organism is unicellular.

C The organism's cells do not contain organelles.

D The organism's cells do not contain genetic information.

---

3. If you were to conduct an investigation to determine if an organism is a plant or an animal, which characteristic could be used to distinguish between the two?

A whether the organism is unicellular or multicellular

B whether or not the organism is made of cells

C whether or not the organism responds to its environment

D whether the organism makes its own food or takes in food

---

4. Which should NOT be included in a model developed to show differences between unicellular and multicellular organisms?

A Unicellular organisms have fewer cells than multicellular organisms.

B Unicellular organisms are larger than multicellular organisms.

C Unicellular organisms are organized differently than multicellular organisms.

D Unicellular organisms are smaller than multicellular organisms.

Real-World Connection

5. **Construct an Argument** Your friend thinks that studying cells is a waste of time. He says, "Cells are so small, most of them can't even be seen without a microscope. So why waste your time focusing on researching things you can't even see?!" Construct an argument detailing the importance of studying cells and how it can affect your friend's life.

6.  **ENGINEERING Connection** Using what you've learned in this lesson, explain how science is dependent on advances in engineering and technology.

Still have questions?
Go online to check your understanding about the characteristics of living things.

Do you still agree with the person you chose at the beginning of the lesson? Return to the Science Probe at the beginning of the lesson. Explain why you agree or disagree with that person now.

EXPLAIN THE PHENOMENON

Revisit your claim about characteristics of living things. Review the evidence you collected. Explain how your evidence supports your claim.

START PLANNING
STEM Module Project
Science Challenge

Now that you've learned about the characteristics of living things, go back to your Module Project to plan your investigation. Remember that you want to determine if the samples perform the same life functions that both you and a unicellular amoeba perform.

PAGE KEELEY SCIENCE PROBES

LESSON 2 LAUNCH

The Basic Unit of Life

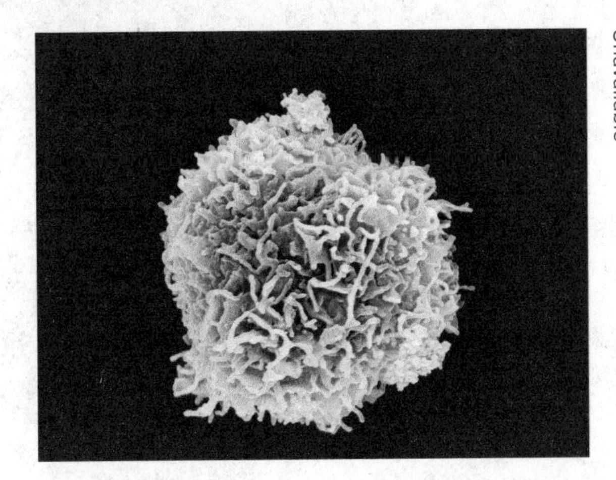

Color-Enhanced SEM Magnification: Unavailable

The cell is called the basic unit of life. What do you think that means? Circle the answer that best matches your thinking.

A. I think it means the cell is the smallest part of matter.

B. I think it means the cell is the smallest part of mass.

C. I think it means the cell is the smallest part of volume.

D. I think it means the cell is the smallest part of mass and volume.

E. I think it means the cell is the smallest part of energy.

F. I think it means the cell is the smallest part of structure.

G. I think it means the cell is the smallest part of structure and function.

H. I think it means the cell is the smallest part of matter, structure, and function.

I. I think it means the cell is the smallest part of matter, energy, and structure.

Explain your answer. Describe your thinking about the cell as a basic unit of life.

You will revisit your response to the Science Probe at the end of the lesson.

LESSON 2

Cell Structure and Function

ENCOUNTER THE PHENOMENON

How do the parts of an ostrich egg—the largest cell on the planet—work together in order for it to function?

Bird eggs have different structures, such as a shell, a membrane, and a yolk. Each structure has a different function that assists in development of the baby bird.

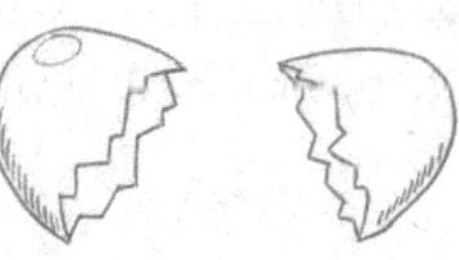

Place an uncooked egg in a bowl. Feel the shell, and record your observations.

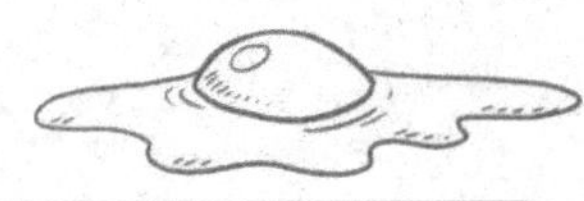

Crack open the egg. Pour the contents into the bowl. Observe the inside of the shell and the contents of the egg. Record your observations.

What do you think is the role of the eggshell? What does the structure of the eggshell tell you about its function?

Eggcellent Science

GO ONLINE Watch the video *Eggcellent Science* to see this phenomenon in action.

EXPLAIN THE PHENOMENON

You observed an egg and inferred the function of its shell. Like the egg, other cells also have separate parts with particular functions. Use your observations about the phenomenon to make a claim about how different parts of a cell work together in order for a whole cell to function.

CLAIM

Parts of a cell work together...

COLLECT EVIDENCE as you work through the lesson.

Then return to these pages to record your evidence.

EVIDENCE

A. What evidence have you discovered to explain what structures surround a cell, such as the ostrich egg, and how these structures help a cell function?

B. What evidence have you discovered to explain what powers a cell?

MORE EVIDENCE

C. What evidence have you discovered to explain what controls a cell?

When you are finished with the lesson, review your evidence. If necessary, based on the evidence, revise your claim.

REVISED CLAIM

Parts of a cell work together...

Finally, explain your reasoning for how and why your evidence supports your claim.

REASONING

The evidence I collected supports my claim because...

What surrounds a cell?

Although different types of cells perform different functions, all cells have some structures in common. Every cell is surrounded by a protective boundary called a membrane. The **cell membrane** is a flexible covering that protects the inside of a cell from the environment outside a cell. What else does the cell membrane do? Let's investigate!

Investigating Cell Membranes

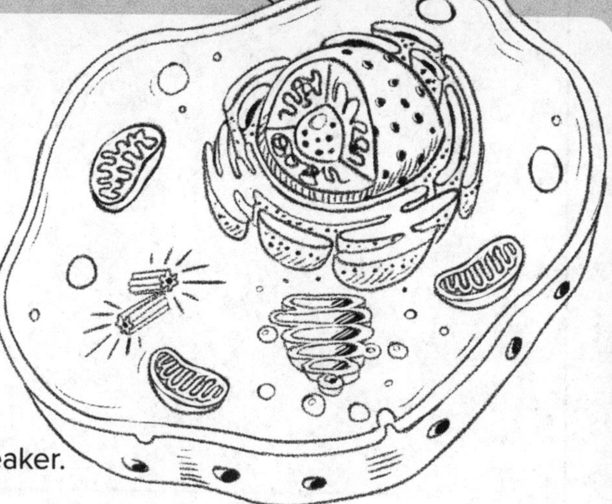

Safety    

Materials

piece of wire mesh
birdseed
500-mL beaker

Procedure

1. Read and complete a lab safety form.
2. Place a square of wire mesh on top of a beaker.
3. Pour a small amount of birdseed on top of the wire mesh. Record your observations in your Science Notebook.
4. Follow your teacher's instructions for proper cleanup.

Analyze and Conclude

5. If the wire mesh in this model represents the cell membrane, how do you think the cell membrane controls what materials enter and leave a cell?

Cell Membrane The cell membrane surrounds the cytoplasm. The **cytoplasm** is a fluid inside a cell that contains salts and other molecules. However, as you just observed, another important role of cell membranes is to control the movement of substances into and out of cells. A cell membrane is semipermeable. This means it allows only certain substances, like nutrients and wastes, to enter or leave a cell.

GO ONLINE for additional opportunities to explore!

Investigate cell membranes by performing one of the following activities.

☐ **Discover** the function of a cell membrane in the **Lab** *How is a balloon like a cell membrane?*

OR

☐ **Explore** more about the cell membrane in the **PhET Interactive Simulation** *Membrane Channels.*

Cell Wall Every cell has a cell membrane, but some cells are also surrounded by a structure called the cell wall. Plant cells, fungal cells, bacteria, and some types of protists have cell walls. Take a look at the image of the plant cell to the right and compare it to the image of the cell on the previous page.

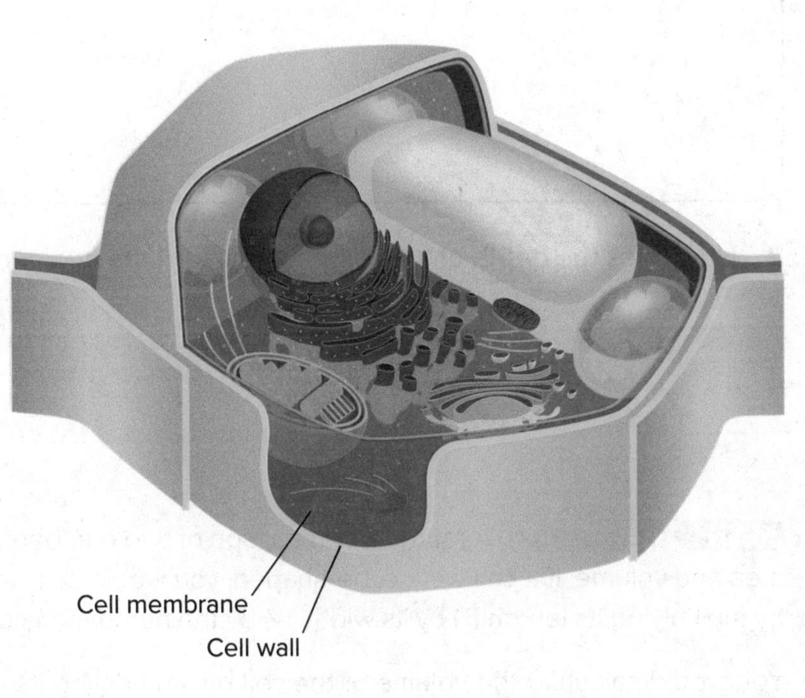

A **cell wall** is a stiff structure outside the cell membrane. A cell wall protects a cell from attack by viruses and other harmful organisms. In some plant cells and fungal cells, a cell wall helps maintain the cell's shape and gives structural support.

COLLECT EVIDENCE

What structures surround a cell, and how do these structures help a cell function? Record your evidence (A) in the chart at the beginning of the lesson.

Want more information?
Go online to read more about cell structure and function.

FOLDABLES®
Go to the Foldables® library to make a Foldable® that will help you take notes while reading this lesson.

How does cell size affect the transport of materials?

Nutrients, oxygen, and other materials enter and leave a cell through the cell membrane. Does the size of a cell affect the transport of these materials throughout the cell? Let's investigate!

INVESTIGATION

Cell Size and Transport of Materials

Watch the demonstration.

1. Use colored pencils to illustrate the two samples you observed in the demonstration.

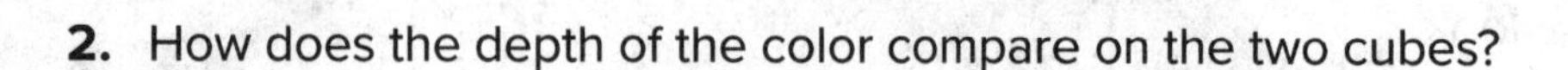

2. How does the depth of the color compare on the two cubes?

MATH Connection A ratio is a comparison of two numbers, such as surface area and volume. If a cell were cube-shaped, you would calculate surface area by multiplying its length (ℓ) by its width (w) by the number of sides (6).

You would calculate the volume of the cell by multiplying its length (ℓ) by its width (w) by its height (h). To find the surface-area-to-volume ratio of the cell, divide its surface area by its volume.

$$\text{Surface area} = \ell \times w \times 6$$

$$\text{Volume} = \ell \times w \times h$$

$$\frac{\text{Surface area}}{\text{Volume}}$$

In the table on the next page, surface-area-to-volume ratios are calculated for cells that are 1 mm and 4 mm per side. Notice how the ratios change as the cell's size increases.

3. Fill in the missing parts of the table in the blue boxes below.

	1 mm × 1 mm × 1 mm cube	4 mm × 4 mm × 4 mm cube
Length	1 mm	4 mm
Width	1 mm	4 mm
Height	1 mm	4 mm
Number of Sides	6	☐
Surface Area ($\ell \times w \times$ **no. of sides**)	1 mm × 1 mm × 6 = ☐ mm^2	4 mm × 4 mm × 6 = 96 mm^2
Volume ($\ell \times w \times h$)	1 mm × 1 mm × 1 mm = 1 mm^3	4 mm × 4 mm × 4 mm = ☐ mm^3
Surface-area-to-volume ratio	$\frac{6\ mm^2}{1\ mm^3} = \frac{6}{1}$ or ☐ :1	$\frac{96\ mm^2}{64\ mm^3} = \frac{1.5}{1}$ or 1.5:1

4. Would a cell with a small surface-area-to-volume ratio be able to transport nutrients and waste through the cell as efficiently as a cell with a large surface-area-to-volume ratio? Explain why or why not.

Surface Area and Volume The movement of nutrients, waste material, and other substances into and out of a cell is important for survival. For this movement to happen, the area of the cell membrane must be large compared to its volume. The area of the cell membrane is the cell's surface area. The volume is the amount of space inside the cell. As a cell grows, both its volume and its surface area increase. The volume of a cell increases faster than its surface area. If a cell were to keep growing, it would need large amounts of nutrients and would produce large amounts of waste material. However, the surface area of the cell's membrane would be too small to move enough nutrients and wastes through it for the cell to survive.

What organelles are involved in the transport of materials?

As you just discovered, the cell membrane enables materials to enter and leave the cell. There are several other organelles related to the transport of materials as well.

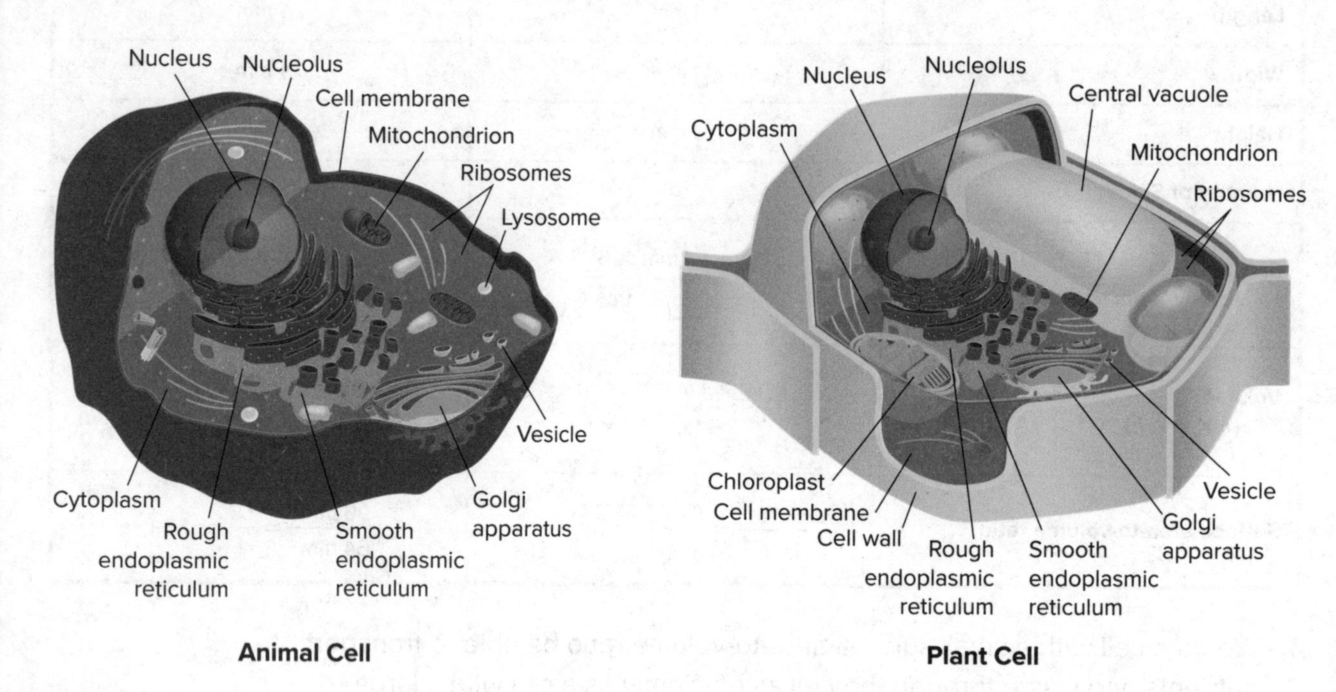

Animal Cell **Plant Cell**

Ribosomes Amino acid molecules made up of carbon, hydrogen, oxygen, nitrogen, and sometimes sulfur, join together to form long chains called **proteins**. Some proteins help cells communicate with each other while others transport substances inside cells. Proteins are made on small structures called ribosomes. Unlike other cell organelles, a ribosome is not surrounded by a membrane.

Endoplasmic Reticulum Ribosomes can be attached to a web like organelle called the endoplasmic reticulum (en duh PLAZ mihk • rih TIHK yuh lum), or ER. The ER spreads from the nucleus throughout most of the cytoplasm. Endoplasmic reticulum with ribosomes on its surface is called rough endoplasmic reticulum. Rough ER is the site of protein production. ER without ribosomes is called smooth ER. Smooth ER is important because it helps remove harmful substances from a cell.

Vacuoles Some cells also have saclike structures called vacuoles. Vacuoles are organelles that store food, water, and waste material. A typical plant cell usually has one large vacuole. Some animal cells have many small vacuoles. A plant cell's vacuole that may take up half of the cell's size. This vacuole not only stores water and other substances, but also enables the plant to stay rigid and supported when filled with water.

The Golgi Apparatus Proteins are prepared for their specific jobs or functions by an organelle called the Golgi apparatus. Then the Golgi apparatus packages the proteins into tiny, membrane-bound, ball-like structures called vesicles. Vesicles are organelles that transport substances from one area of a cell to another area of a cell. Some vesicles in an animal cell are called lysosomes. Lysosomes contain substances that help break down and recycle cellular components.

THREE-DIMENSIONAL THINKING

Create a graphic organizer that explains how the various **structures**, or organelles, you just learned about help a cell **function** as a whole.

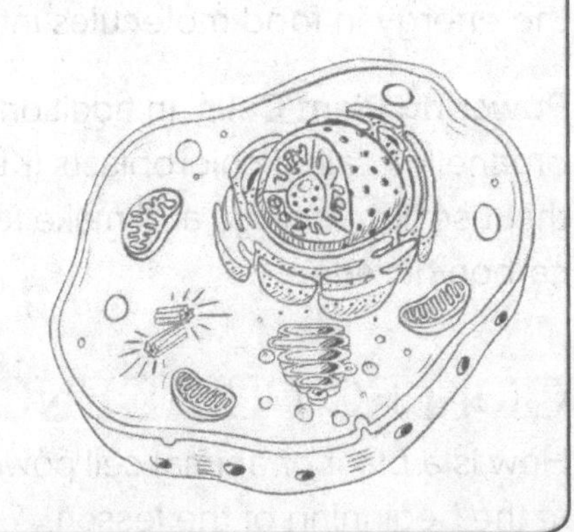

What powers cellular activity?

You learned about how cells transport materials across the cell membrane. How does the cell power such complex activity?

INVESTIGATION

Powering a Cell

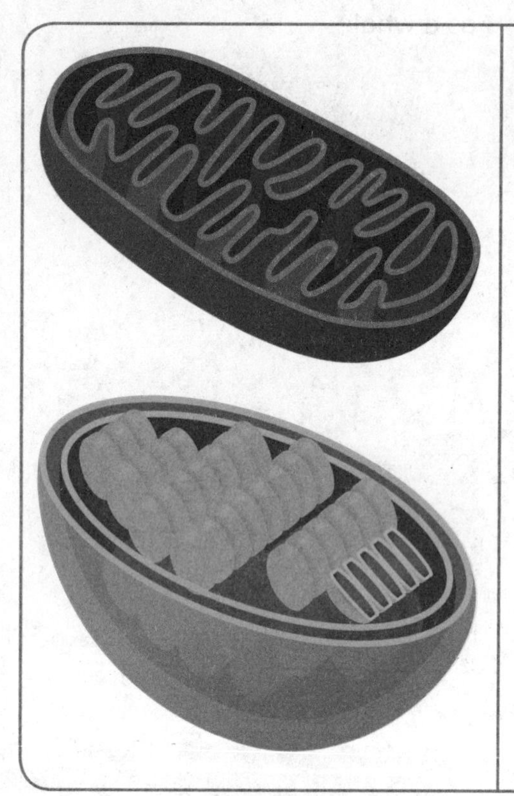

What do you notice about the two organelles to the left? What are their differences, and what are their similarities? Can you infer what their functions are?

Mitochondria The bean-shaped organelle on top is called a mitochondrion, and it powers the cell through chemical reactions. Mitochondria are found in both plant and animal cells. It has two membranes to increase the surface area for these reactions to occur. Mitochondria are a vital part of cellular respiration. **Cellular respiration** is a series of chemical reactions that convert the energy in food molecules into a usable form of energy called ATP.

Powering Plant Cells In addition to mitochondria, plant cells contain organelles called chloroplasts (KLOR uh plasts). **Chloroplasts** are organelles that use light energy and make food—a sugar called glucose—from water and carbon dioxide.

COLLECT EVIDENCE

How is a plant or animal cell powered? Record your evidence (B) in the chart at the beginning of the lesson.

What controls all of this activity?

The largest organelle inside most eukaryotic cells is the nucleus. The **nucleus** is the part of a eukaryotic cell that directs cell activities and contains important cellular information stored in DNA. DNA is organized into structures called chromosomes. The DNA of each cell carries information that provides instructions for making all the proteins a cell requires.

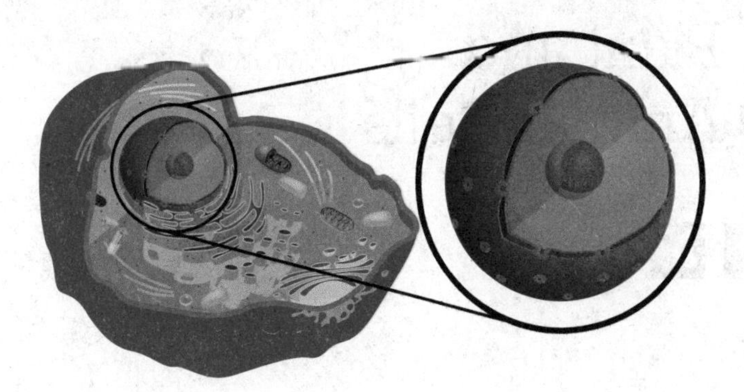

In addition to chromosomes, the nucleus contains proteins and an organelle called the nucleolus (new KLEE uh lus). The nucleolus makes ribosomes, organelles that are involved in the production of proteins. The nucleus controls all cell activity by directing protein synthesis. Proteins are needed for almost every function in the body.

THREE-DIMENSIONAL THINKING

Can you think of some analogies for the nucleus? Use the space below to make a drawing, diagram, or other illustration to help you understand the structure and function of the nucleus.

COLLECT EVIDENCE

What controls a cell? Record your evidence (C) in the chart at the beginning of the lesson.

What is the difference between plant and animal cells?

You've learned that plants and animal cells differ in that plant cells have cell walls, but animal cells don't. How else are plant and animals cells different?

LAB Plant and Animal Cells

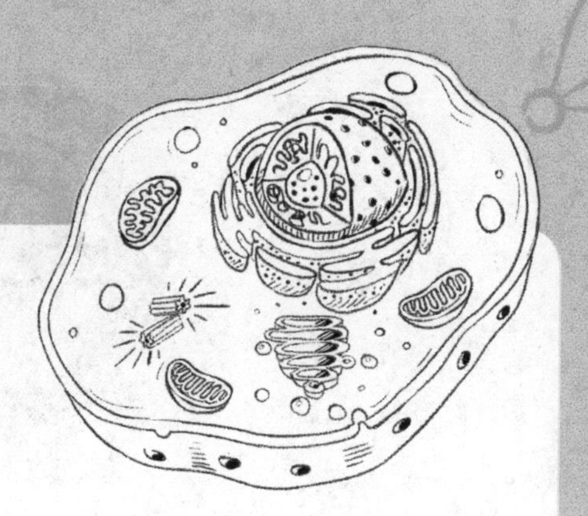

Safety     

Materials

microscope

microscope slide and coverslip

tweezers or forceps

dropper

Elodea plant

prepared slide of human cheek cells

Procedure

1. Read and complete a lab safety form.
2. Using tweezers or forceps, make a wet mount slide of a young leaf from the tip of an *Elodea* plant.
3. Use a microscope to observe the leaf on low power. Focus on the top layer of cells.
4. Switch to high power and focus on one cell. Moving around the central vacuole are green, disklike objects called chloroplasts. Try to find the nucleus. It looks like a clear ball.
5. Draw a diagram of an *Elodea* cell in the Data and Observations section. Label the cell wall, chloroplasts, cytoplasm, and nucleus. Return to low power and remove the slide. Properly dispose of the slide.
6. Observe the prepared slide of cheek cells under low power.
7. Switch to high power and focus on one cell. Draw a diagram of one cheek cell in the Data and Observations section. Label the cell membrane, cytoplasm, and nucleus. Return to low power and remove the slide.
8. Follow your teacher's instructions for proper cleanup.

Data and Observations

Analyze and Conclude

9. Based on your diagrams, how do the shapes of the *Elodea* cell and cheek cell compare?

10. Compare and contrast the cell structures in your two diagrams. Which structures did you observe in both cells? Which structures did you observe in only one of the cells?

What can different cells do?

You might recall from Lesson 1 that all living things are made up of one or more cells. Multicellular organisms have different types of cells with different functions that enable the survival of the entire organism. Cells come in many shapes and sizes. Explore how the structure of a cell relates to what it does.

THREE-DIMENSIONAL THINKING

The **structure** of a cell relates to its job, or **function.** Use the table below to infer a cell's function based on its shape.

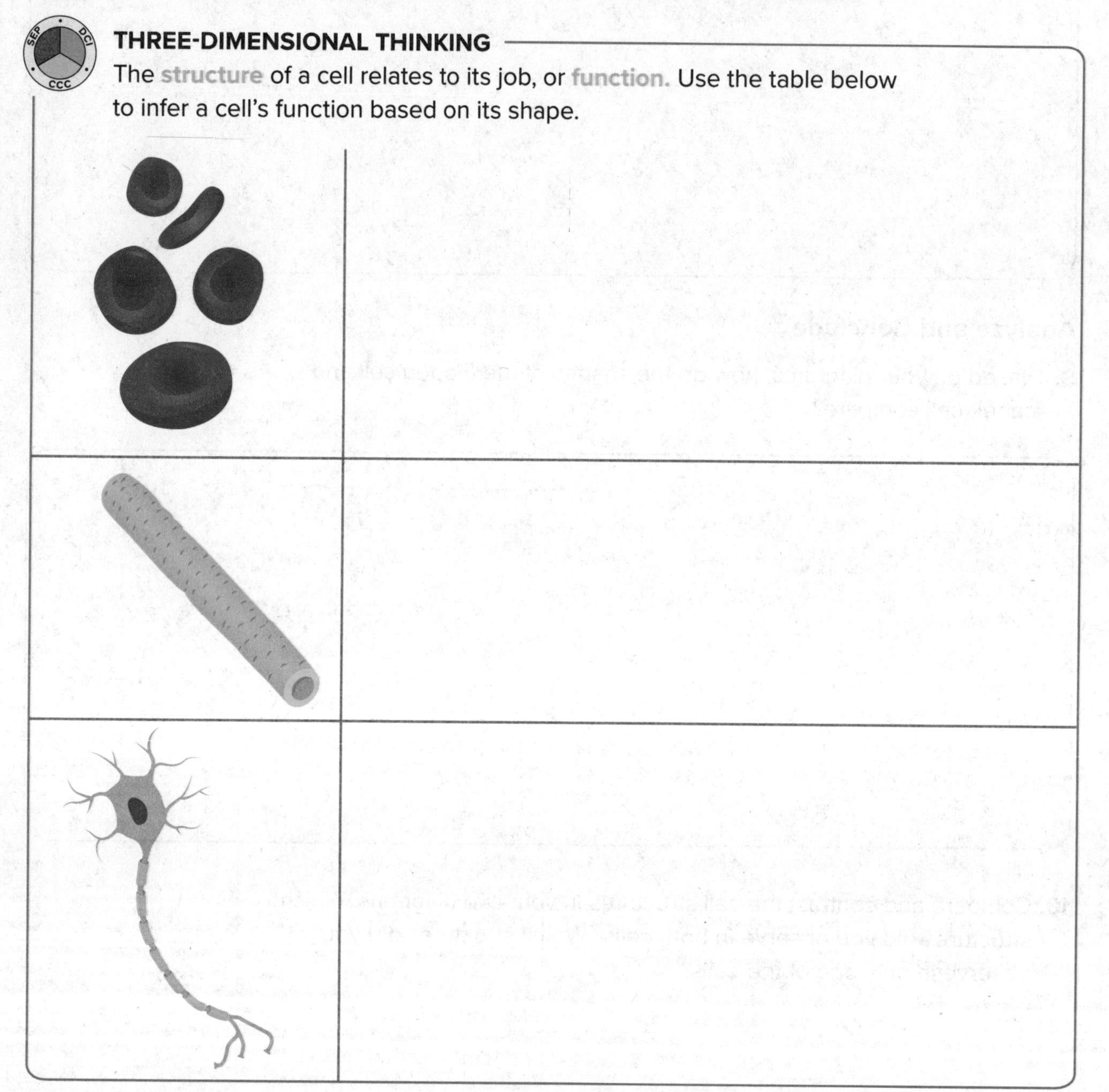

Types of Cells and Structures Cells in the body can be incredibly diverse, as you just saw. Red blood cells are disk-shaped, which helps them move through blood vessels so that they can carry oxygen throughout the body. Xylem cells are tubelike cells that transport water from the roots to the leaves of plants. The neuron is a cell found in many animals that transmits impulses from different parts of the body. Each cell is unique but works with other cells as body functions are carried out.

A Closer Look: Vivacious *Vorticella*

Color-Enhanced SEM
Magnification: 1,300×

The organism you see isn't an alien but a unicellular organism called a *Vorticella*! The *Vorticella* is a protozoan found in freshwater ponds and lakes. The *Vorticella* may be small, but it has everything it needs to survive. The opening of the cell may look like it is full of hair, but the structures are actually appendages that help it gather food. The hairlike appendages are called cilia, and they create small water currents that bring food toward the cell. And that spiral appendage isn't a tail; it's a stalk that allows the cell to latch onto surfaces.

HISTORY Connection *Vorticella* was first discovered by a Dutch scientist named Anton van Leeuwenhoek who thought its mouth parts were horns. Since then, scientists studying the organism have discovered its possibilities as a pest controller. *Vorticella* is able to bind to mosquito larvae and prevent them from reaching maturity. Mosquitos are known to carry pathogens that are dangerous to humans.

Vorticella may be small, but it has everything it needs to survive and thrive.

LM Magnification: 430×

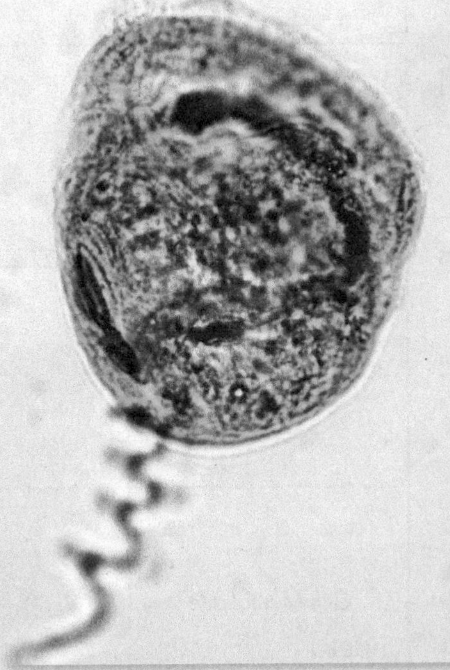

It's Your Turn

READING Connection With a partner, choose another unicellular organism to research. Create a presentation with a visual aid that details how the organism's structures relate to functions that ensure its survival.

LESSON 2

Review

Summarize It!

1. **Classify** information about organelles and cell structures. In the right-hand column, indicate whether the organelle is in a plant cell, an animal cell, or both.

Organelle	*Function*	*Plant, animal, or both?*
Nucleus		
Mitochondria		
Chloroplasts		
Cell Wall		
Cell Membrane		

Three-Dimensional Thinking

2. Rosa is planning an investigation using a microscope to try to identify a group of cells. She sees that the cells are joined together, so she knows that they are from one organism. If she also sees that all of the cells have cells walls, Rosa can conclude that she could be looking at

A bacterial cells.

B human cells.

C mouse cells.

D plant cells.

---

3. Mitochondria function as subsystems within the larger system of the cell as a whole. Which explains why a mitochondrion, shown below, is known as the "power house" of a cell?

A It converts energy in food to ATP.

B It helps the cell gather sunlight.

C The cell eats it as food.

D It has two membranes.

---

4. **MATH Connection** Which statement could you use to construct an explanation for why it is important for a cell's surface-area-to-volume ratio to not be too small?

A Wastes and nutrients need to move through the membrane.

B If a cell's surface-area-to-volume ratio was too small, the cell would starve.

C If a cell's surface-area-to-volume ratio was too small, the cell would not produce enough waste material.

D If a cell's surface-area-to-volume ratio was too small, the organelles would grow too large to fit within the cell.

Real-World Connection

5. **Infer** Suppose that you are a scientist and you have been given a sample of unknown cells. By looking at the cells under a microscope, what would you be able to determine about the organisms the cells came from? Explain your reasoning.

6. **Explain** Your friend is making a model of a cell and wants to use metal to represent the cell membrane because metal is solid and would allow nothing to enter or leave the cell. Explain why you agree or disagree with his reasoning.

Still have questions?
Go online to check your understanding about cell structure and function.

REVISIT

Do you still agree with the answer you chose at the beginning of the lesson? Return to the Science Probe at the beginning of the lesson. Explain why you agree or disagree with that answer now.

EXPLAIN THE PHENOMENON

Revisit your claim about how the parts of a cell contribute to the function as a whole. Review the evidence you collected. Explain how your evidence supports your claim.

PLAN AND PRESENT

STEM Module Project
Science Challenge

Now that you've learned about structures and functions of cells, go back to your Module Project to build your model and present to the panel. Keep in mind that you want your model to show how cell structures allow organisms, such as the amoeba, to perform life functions.

PERFORMANCE EXPECTATION

IT'S ALIVE! Or is it?

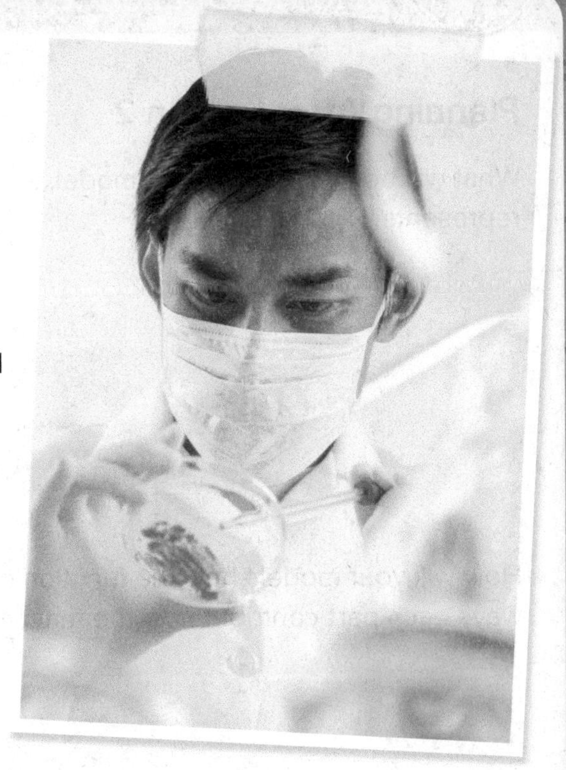

You have been invited by scientists to work on a special project involving the possibility of life on another planet. Your job is to help decide whether or not life has been found in outer space.

Astronauts were able to travel to a nearby planet and gather samples. Some of the samples appear to be living, but the scientists don't know for sure.

Your goal is to determine whether or not the samples are living by conducting an investigation to show what living things are made of, and developing and using a model of the building blocks of a living thing to show how all living things, from yourself to the microscopic amoeba at the beginning of the module, perform functions to stay alive. You will present your investigation and model to a panel of scientists.

Planning After Lesson 1

List the steps in your investigation to show whether the samples are living or nonliving.

How will scientists conclude whether the samples from the other planet are living or nonliving?

Planning After Lesson 2

What will be the parts in your model, and how will you show what each part represents?

How will your model show the function of the system as a whole and the ways each part contributes to the function as a whole?

How will your model help the scientists determine whether the samples are living or nonliving?

Look at the planning you did after each lesson. Use that information to complete the plan for your investigation in your Science Notebook, and build your model.

Explain Your Investigation and Use Your Model

Now that you've planned your investigation and developed your model, complete the tables below.

Investigation	
Purpose What is the phenomenon under investigation?	
Evidence What data will be collected to address the purpose of the investigation?	
Planning Explain the methods and tools to be used in the investigation. Keep in mind the size and scale of cells.	

Model	
Components What are the different parts of my model?	
Relationships How do the components of my model interact?	
Connections How does my model help me understand the phenomenon?	

Give Your Presentation

Analyze and evaluate your plan and model before you make your presentation for the panel of scientists.

How do your investigation and model help you better understand how the amoeba at the beginning of the module is a living thing that performs the same functions that you do in order to stay alive?

Congratulations! You've completed the Science Challenge requirements.

Module Wrap-Up

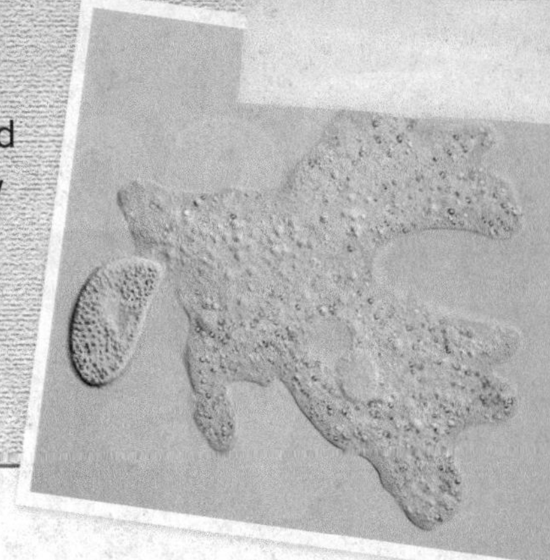

REVISIT THE PHENOMENON

Using the concepts you have learned throughout this module, explain how the amoeba shown at the beginning of the module performs all the same functions that you do to stay alive.

OPEN INQUIRY

What are one or two questions you still have about the phenomenon?

Choose the question that interests you the most. Plan and conduct an investigation to answer this question.

Body Systems

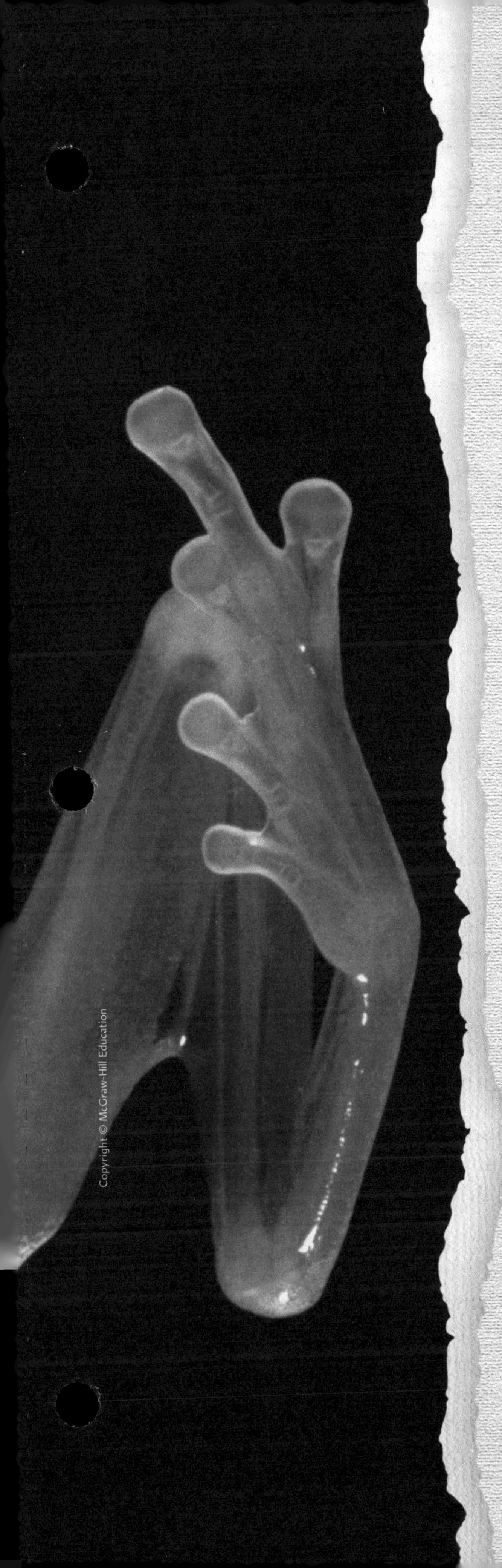

ENCOUNTER THE PHENOMENON

How do body systems in this glass frog work together to perform life functions?

Glass Frog

GO ONLINE
Watch the video *Glass Frog* to see this phenomenon in action.

Collaborate With your class, develop a list of questions that you could investigate to find out more about how the glass frog's body systems work together to perform life functions. Record your questions below.

STEM Module Project Launch
Science Challenge

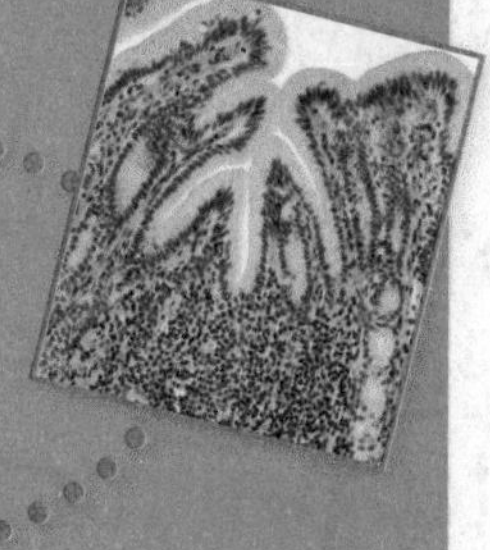

Lesson 1
Levels of Organization

Lesson 2
Structure and Support

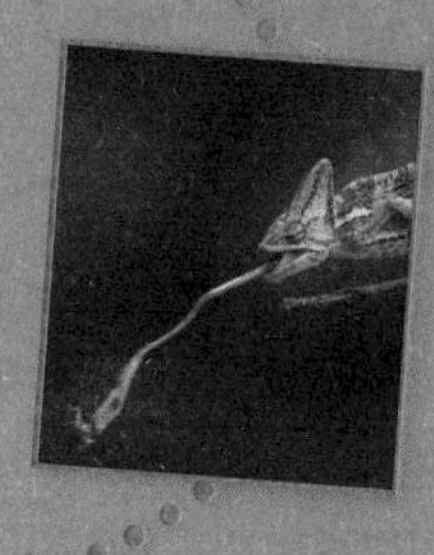

Lesson 3
Obtaining Energy and Removing Waste

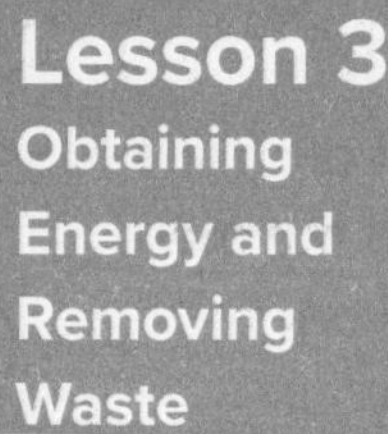

Lesson 4
Moving Materials

Lesson 5
Control and Information Processing

Body of Evidence

"Hey, Mr. Fernandez! We won our soccer game, thanks to my super strong muscles! I scored the winning goal!"

"That's great, Anna, but you know that you need more than your muscles to play soccer, right?"

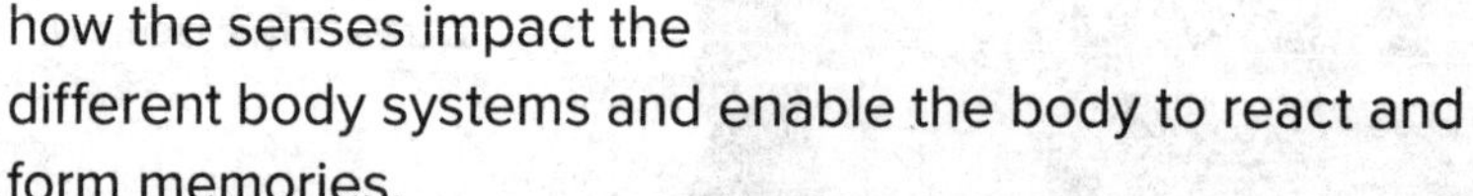

Your team's task is to prepare to debate your classmate, who thinks that the body is made of independent subsystems that do not interact. You must provide evidence to support your argument that body systems in organisms, such as the soccer player and the glass frog on the previous pages, interact, as well as information about how the senses impact the different body systems and enable the body to react and form memories.

Start Thinking About It

In the photo above, you see a girl playing soccer. What body parts or body systems do you think the girl is using in the photo? Discuss your thoughts with your group.

STEM Module Project

Planning and Completing the Science Challenge
How will you meet this goal? The concepts you will learn throughout this module will help you plan and complete the Science Challenge. Just follow the prompts at the end of each lesson!

LESSON 1 LAUNCH

PAGE KEELEY SCIENCE PROBES

Basic Unit of Function

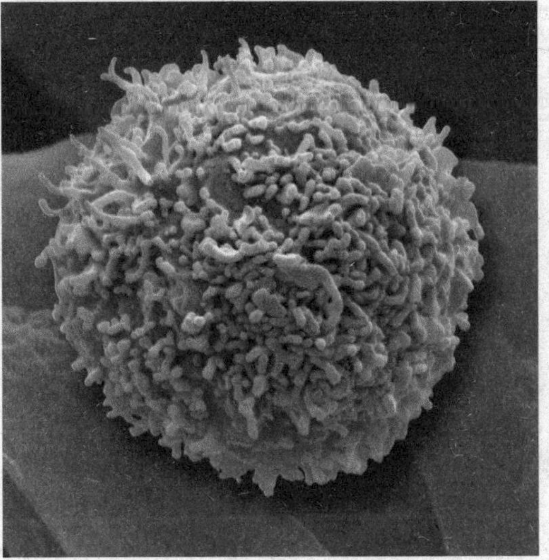

Color-Enhanced SEM Magnification: 6,000×

Four friends talked about the human body. They each agreed that the cell is the basic unit of structure in the human body. But they did not agree on the basic unit of function in the human body. This is what they said:

Sarah: I think body systems are the basic unit of function.

Todd: I think cells are the basic unit of function in the body.

Juanita: I think tissues are the basic unit of function.

Seif: I think organs are the basic unit of function.

Circle the friend you most agree with. Explain your thinking.

You will revisit your response to the Science Probe at the end of the lesson.

LESSON 1

Levels of Organization

ENCOUNTER THE PHENOMENON | How are your cells organized to make up your body?

GO ONLINE
Watch the video *Pointillism* to see this phenomenon in action.

The cells in the photo appear to be grouped together. Pointillism is a style of painting in which many tiny dots are grouped in ways to form an image. How might this be similar to the way your body is organized?

Scientists estimate there are 37.2 trillion cells in your body. All of those cells are organized in a way that enables your body to function. With a partner, use markers or colored pencils in the space below to create a basic piece of pointillism art. Then explain how this style of art might be similar to the way the cells in your body are organized.

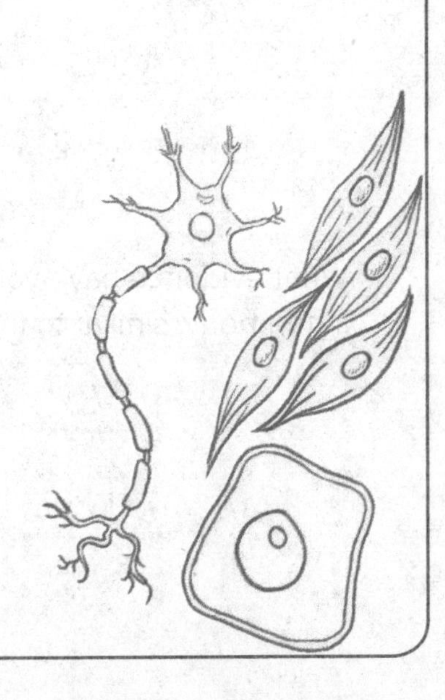

EXPLAIN THE PHENOMENON

Did you see how the dots in the art grouped together to form an image? Use your observations about the phenomenon to make a claim about how cells are organized to make up organisms.

CLAIM

Cells are organized...

COLLECT EVIDENCE as you work through the lesson.

Then return to these pages to record your evidence.

EVIDENCE

A. What evidence have you discovered to explain how cells are organized in the body, similar to the way dots in pointillism art are organized?

B. What evidence have you discovered to explain how tissues are organized in the body, similar to the way dots in pointillism art are organized?

MORE EVIDENCE

C. What evidence have you discovered to explain how organs are organized in the body, similar to the way dots in pointillism art are organized?

D. What evidence have you discovered to explain how organ systems are organized in the body, similar to the way dots in pointillism art are organized?

When you are finished with the lesson, review your evidence. If necessary, based on the evidence, revise your claim.

REVISED CLAIM

Cells are organized...

Finally, explain your reasoning for how and why your evidence supports your claim.

REASONING

The evidence I collected supports my claim because...

How are cells organized in the body?

Organisms can be unicellular—made of one cell, or multicellular—made of more than one cell. You've learned that your own body has trillions of cells! Multicellular organisms have different types of cells that each perform a specific job.

As multicellular organisms grow, cells divide to produce new cells. The first cells made can become any type of cell, such as a muscle cell, a nerve cell, or a blood cell, through the process of **cell differentiation.** As the number of cells in an organism increases, similar types of cells are organized into groups.

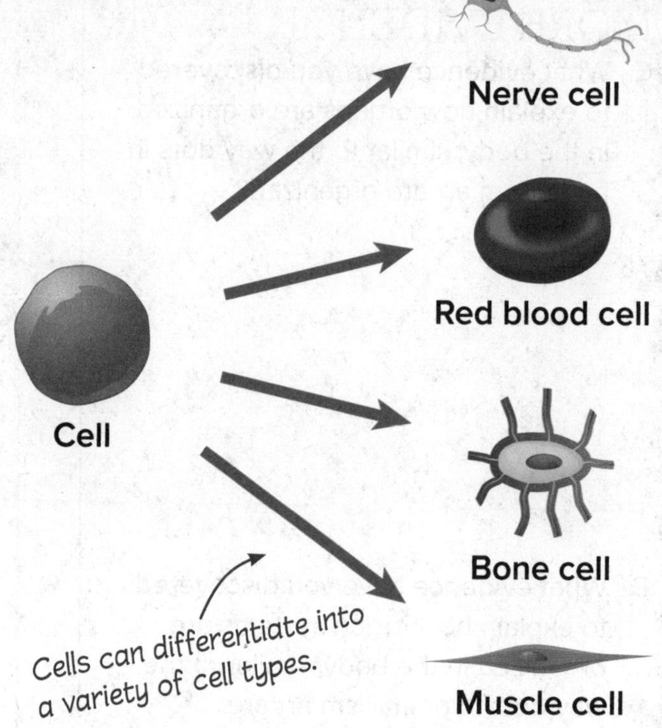

INVESTIGATION

When Cells Get Together

1. Look at the cells provided by your teacher. What type of cell do you think each is? What do you think each cell does?

2. What larger part of the body do you think cells make up?

Want more information?
Go online to read more about levels of organization in organisms.

FOLDABLES
Go to the Foldables® library to make a Foldable® that will help you take notes while reading this lesson.

Tissues Cells group together to form tissues. **Tissues** are groups of similar types of cells that work together to carry out specific tasks. Humans, like most other animals, have four main types of tissue—muscle, connective, nervous, and epithelial (eh puh THEE lee ul). Muscle tissue, shown in the photo to the right, causes movement. Connective tissue provides structure and support and often connects other types of tissue together. Nervous tissue carries messages to and from the brain. Epithelial tissue forms the protective outer layer of the skin and the internal lining of the body.

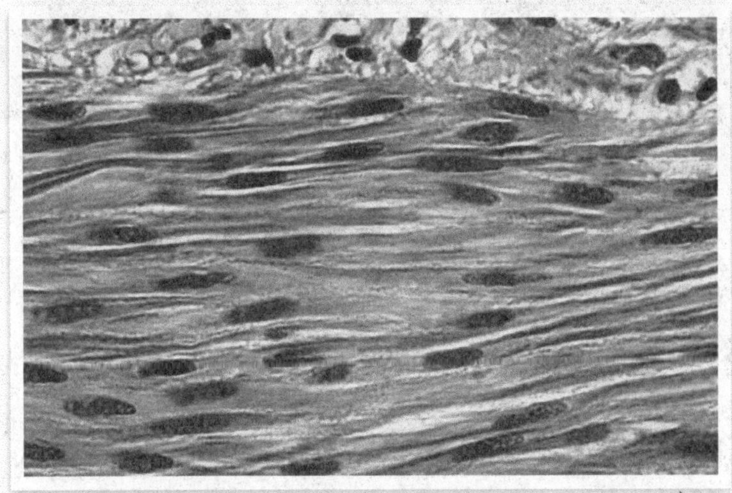

LM Magnification: 250×

This muscle tissue contracts the stomach to help digestion.

What do you think these dark spots are?

Plants also have different types of tissues. The three main types of plant tissue are dermal, vascular (VAS kyuh lur), and ground tissue. Dermal tissue provides protection and helps reduce water loss. Vascular tissue, shown in the photo to the right, transports water and nutrients from one part of a plant to another. Ground tissue provides storage and support and is where photosynthesis takes place.

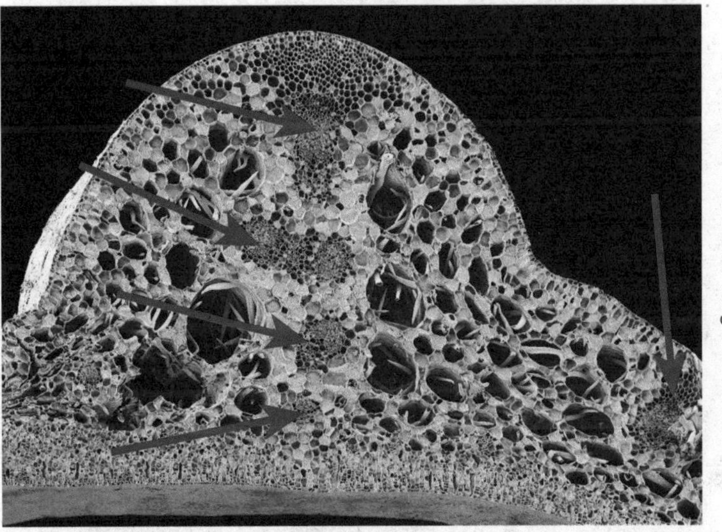

Color-Enhanced SEM Magnification: 30×

Plant vascular tissue, indicated by arrows, moves water and nutrients throughout a plant.

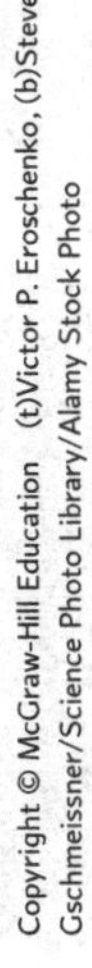

THREE-DIMENSIONAL THINKING

WRITING Connection Write an **argument** that **explains** which human tissue is most similar in **function** to dermal tissue in plants.

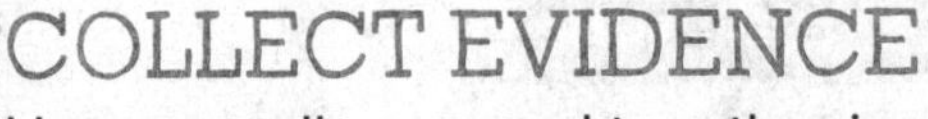

COLLECT EVIDENCE

How are cells grouped together in an organism to form a larger system within the body, similar to the way dots in pointillism art are organized? Record your evidence (A) in the chart at the beginning of the lesson.

SCIENCE & SOCIETY

Bone Marrow Transplants

Why might you need new bone marrow?

▲ In healthy bone marrow, a stem cell can develop into different types of blood cells.

Healthy blood cells are essential to overall health. Red blood cells carry oxygen throughout the body. Some white blood cells fight infections. Platelets help stop bleeding. A bone marrow transplant is sometimes necessary when a disease interferes with the body's ability to produce healthy blood cells.

Bone marrow is a tissue found inside some of the bones in your body. Healthy bone marrow contains cells that can develop into white blood cells, red blood cells, or platelets. Some diseases, such as leukemia and sickle cell disease, affect bone marrow. Replacing malfunctioning bone marrow with healthy bone marrow can help treat these diseases.

A bone marrow transplant involves several steps. The patient receiving the bone marrow must have treatments to destroy his or her unhealthy bone marrow. Healthy bone marrow must be obtained for the transplant. Sometimes, the patient's own bone marrow can be treated and used for transplant. This transplant has the greatest chance of success. Other transplants involve healthy bone marrow donated by another person. The bone marrow must be tested to ensure that it is a good match for the patient.

The bone marrow donor undergoes a procedure called harvesting. Bone marrow is taken from the donor's pelvic, or hip, bone. The donor's body replaces the harvested bone marrow, so there are no long-term effects for the donor. The donated bone marrow is introduced into the patient's bloodstream. If the transplant is successful, the new bone marrow moves into the bone cavities and begins producing healthy blood cells.

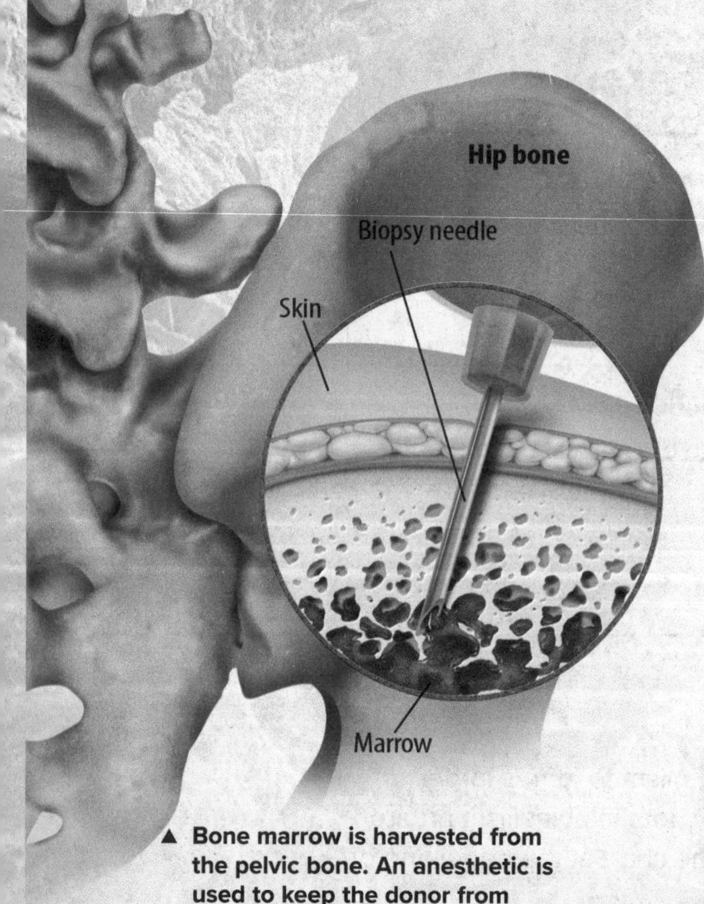

▲ Bone marrow is harvested from the pelvic bone. An anesthetic is used to keep the donor from feeling pain during the procedure.

It's Your Turn

Research and Report Find out more about bone marrow transplants. What other diseases can be treated using a bone marrow transplant? What is the National Marrow Donor Program? Present your findings to your class.

How are tissues organized in the body?

When similar cells are grouped together, they become a tissue. What do you think happens when tissues are organized into groups?

These cells are grouped together to form fat—a connective tissue.

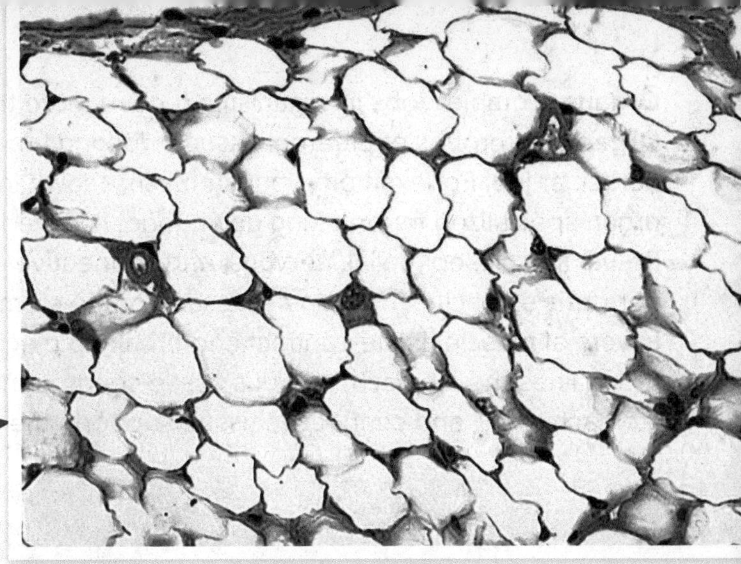

LM Magnification: 200×

INVESTIGATION

Making Bodies

1. Look at the cell card provided by your teacher. Move around the room and meet up with other students to form a tissue—either muscle tissue, connective tissue, nervous tissue, epithelial tissue, dermal tissue, vascular tissue, or ground tissue. What tissue did you make? How did you know which students to meet up with?

2. Now you will form a larger part of either a plant or animal. If you are in an animal tissue group, meet up with all the other animal tissues, and if you are in a plant tissue group, meet up with all other plant tissue groups. Explain what part of the plant or animal you could be making.

3. What larger body part do you think tissues form?

Organs Complex jobs in organisms require more than one type of tissue. **Organs** are groups of different tissues working together to perform a particular job. For example, your stomach, shown in the image below, is an organ specialized for breaking down food. It is made of all four types of tissue: muscle, epithelial, nervous, and connective. Each type of tissue performs a specific function necessary for the stomach to work properly. Layers of muscle tissue contract and break up pieces of food, epithelial tissue lines the stomach, nervous tissue sends signals to indicate the stomach is full, and connective tissue supports the stomach wall.

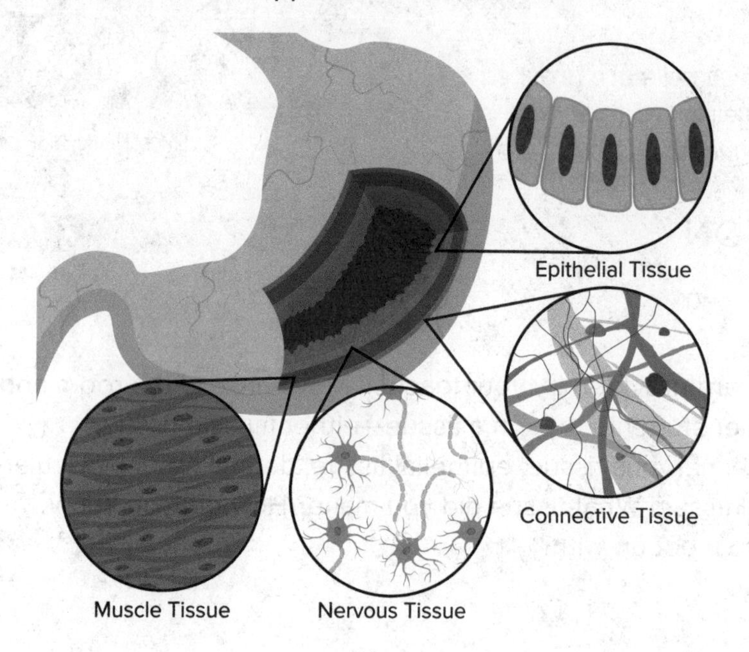

Plants also have organs. Leaves are organs specialized for photosynthesis. Each leaf is made of dermal, ground, and vascular tissues. Dermal tissue covers the outer surface of a leaf. The leaf is a vital organ because it contains ground tissue that produces food for the rest of the plant. Ground tissue is where photosynthesis takes place. The ground tissue is tightly packed on the top half of a leaf. The vascular tissue moves both the food produced by photosynthesis and water throughout the leaf and the rest of the plant.

A plant leaf is an organ made of several different tissues.

COLLECT EVIDENCE

How are tissues grouped together in an organism to form a larger system within the body, similar to the way dots in pointillism art are organized? Record your evidence (B) in the chart at the beginning of the lesson.

How are organs organized in the body?

Each individual organ is important to an organism's survival. What important functions occur when organs are organized into groups?

INVESTIGATION

Body Functions

1. Think about all the different tasks bodies perform to stay alive. Brainstorm functions of the body below.

2. In your group, look at the diagram of the body. Think about the body function assigned to you by your teacher. What organs do you think need to work together to perform that function? How do you think those organs work together to perform that particular function?

Organ Systems Usually organs do not function alone. Instead, **organ systems** are groups of different organs that work together to complete a series of tasks. For example, the human digestive system is made of many organs, including the stomach, the small intestine, the liver, and the large intestine. These organs and others all work together to break down food and take it into the body.

Plants have two major organ systems—the shoot system and the root system. The shoot system includes leaves, stems, and flowers. Food and water are transported throughout the plant by the shoot system. The root system anchors the plant and takes in water and nutrients.

COLLECT EVIDENCE

How are organs grouped together in an organism to form a larger system within the body, similar to the way dots in pointillism art are organized? Record your evidence (C) in the chart at the beginning of the lesson.

How are organ systems organized in the body?

In a multicellular organism, similar cells work together and make a tissue. Tissues are organized into organs, and organs are organized into organ systems which work together to keep an organism functioning. How can you model the levels of organization in an organism?

LAB Organism Organization

Safety 

Materials

cardboard shape
permanent marker
tape
macaroni
glue

Procedure

1. Read and complete a lab safety form.
2. Your teacher will give you a cardboard shape, macaroni, and a permanent marker.

3. The macaroni represents cells. Use the marker to draw a small circle on each piece of macaroni. This represents the nucleus.

4. Arrange and glue enough macaroni on the blank side of the cardboard shape to cover it. Your group of similar cells represents a

    ____________________.

5. One of the squares on the back of your shape is labeled A, B, C, or D. Find the group with a matching letter. Line up these squares, and use tape to connect the two tissues. What does this represent?

    ____

6. Repeat step 5 with the squares labeled E or F. This represents an

    ____________________.

7. Connect the organ systems by aligning the squares labeled G to represent an organism.

8. Follow your teacher's instructions for proper cleanup.

Analyze and Conclude

9. Each group had to work with other groups to make a model of an organism. Do cells, tissues, and organs need to work together in organisms? Explain.

10. How does your model show the levels of organization in living things?

Organisms Multicellular organisms usually have many organ systems. These systems work together to carry out all the jobs needed for the survival of the organisms. For example, the cells in the leaves and the stems of a plant need water to live. They cannot absorb water directly. Water diffuses into the roots and is transported through the stem to the leaves by the shoot, or transport, system.

In the human body, there are many major organ systems. Each organ system depends on the others and cannot work alone. For example, the cells in the muscle tissue of the stomach cannot survive without oxygen. The stomach cannot get oxygen without working together with the respiratory and circulatory systems.

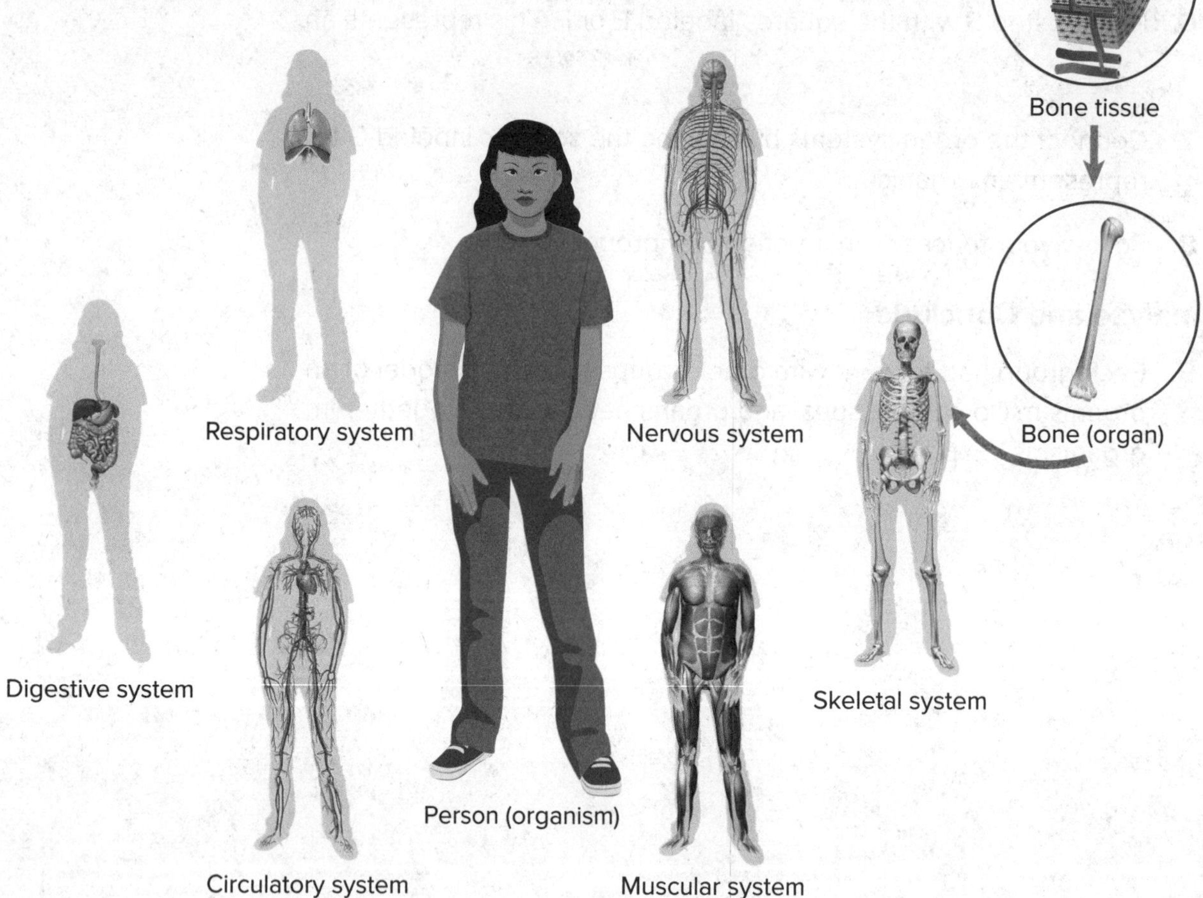

COLLECT EVIDENCE

How are organ systems grouped together to form a larger system, similar to the way dots in pointillism art are organized? Record your evidence (D) in the chart at the beginning of the lesson.

THREE-DIMENSIONAL THINKING

Explain the difference in **scale, proportion, and quantity** in the **systems** and subsystems that make up the levels of organization in organisms. Record your response in your Science Notebook.

A Closer Look: Organ Donation

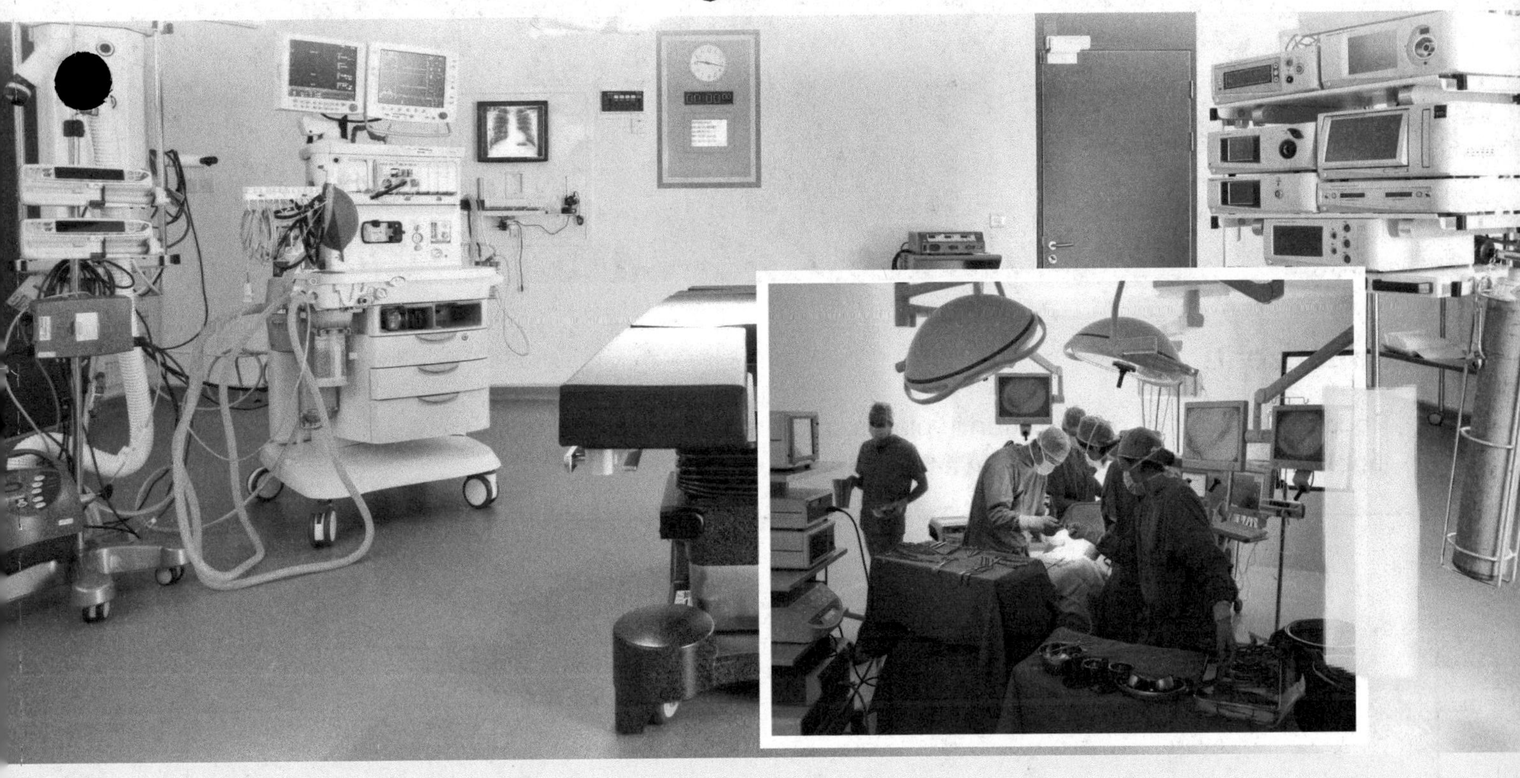

Every 10 minutes someone in the United States is placed on the national transplant waiting list to wait for an organ that could save her life. There are around 120,000 people in the United States that are currently on the transplant list. However, only 54% of Americans are registered as organ donors, and many patients on the waitlist will not receive a donation. An organ donor is a person who has agreed that in the event of his death, his organs can be transplanted to a patient in need.

Some organs, like the kidney or liver, can be donated from a living donor. Both the donor and recipient can go on to live healthy lives. The kidney is the most frequent type of living organ donation. In a kidney donation, the donor would give one of her two healthy kidneys to the recipient. Each would then be able to live with a single kidney. Most people on the national transplant waiting list are in need of a kidney.

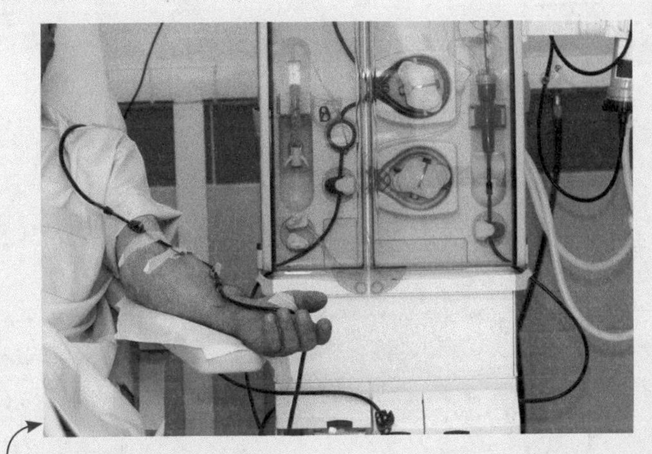

People waiting for a kidney transplant need dialysis—a process that artificially removes wastes from blood.

It's Your Turn

Investigate Interview a teacher, family member, or friend who is a registered organ donor. Create a list of questions to ask them relating to their decision to become an organ donor. Find out why they decided to make this choice.

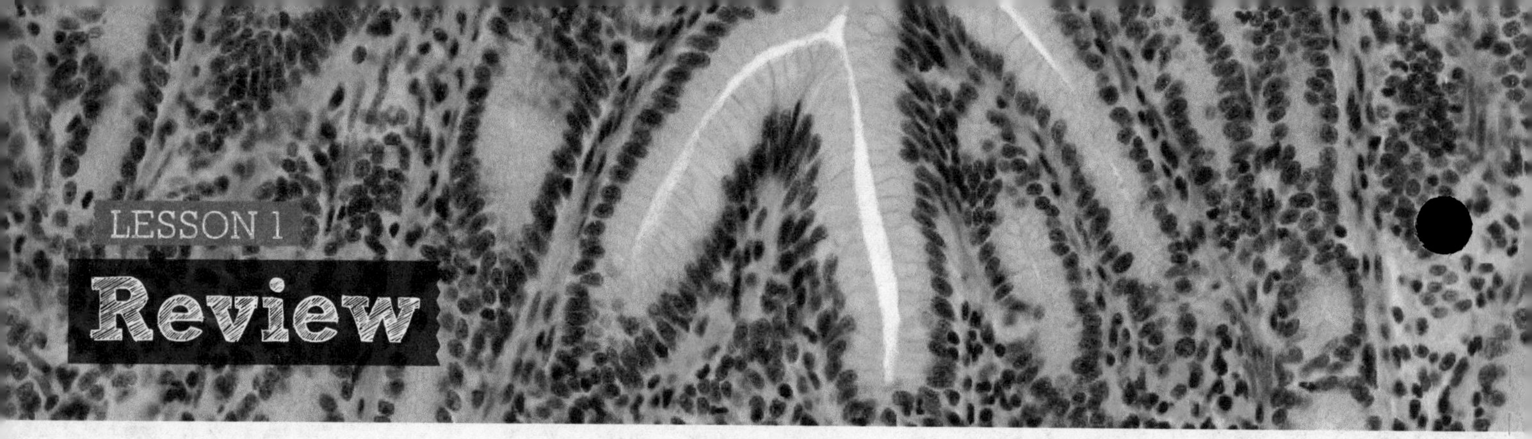

LESSON 1

Review

Summarize It!

1. **Recognize** levels of organization in plants. In the space below draw and label the levels of organization in a tree.

Three-Dimensional Thinking

Cade is making a presentation on the way systems are organized in plants and animals for his science class. He prepared this flowchart to illustrate his presentation.

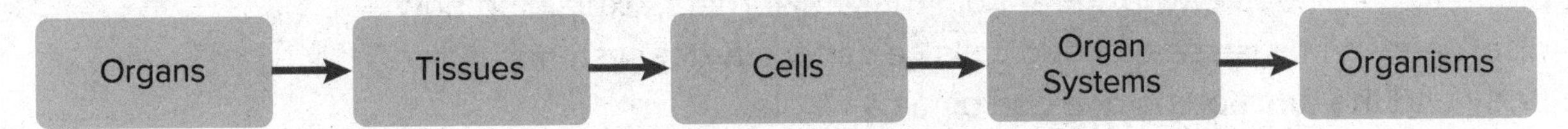

2. After further study, Cade realized his flowchart was incorrect. Which change should he make to correct the flowchart?

A add the phrase *Cell Membranes*

B move *Organ Systems* to the beginning

C remove *Tissues* and *Cells*

D switch *Organs* and *Cells*

The diagram shows the structures involved in respiration for a grasshopper.

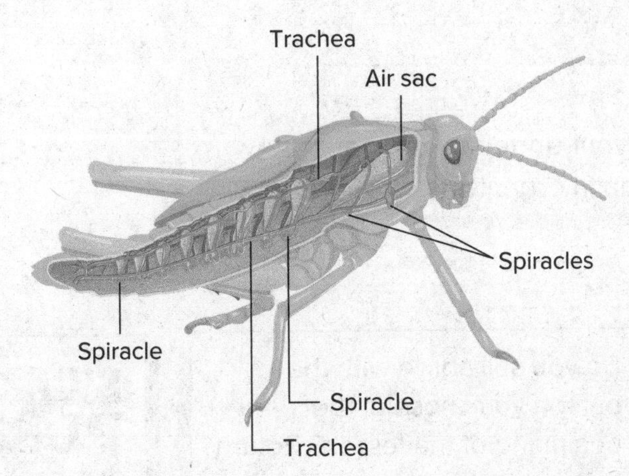

3. These structures are an example of which level of organization within an organism?

A The structures make up an organelle.

B The structures make up an organ system.

C The structures make up a specialized cell.

D The structures make up a tissue.

Real-World Connection

4. Argue There are many different disorders and diseases that affect a person's blood cells. Some of these disorders and diseases require blood transfusions—taking blood from a healthy donor and inserting it into the veins of the person in need. Write an argument in support of donating blood to help people who suffer from these disorders, explaining how disorders and diseases of blood cells can affect larger systems in the body and the functioning of a person as a whole.

Still have questions?
Go online to check your understanding about levels of organization in organisms.

REVISIT

Do you still agree with the person you chose at the beginning of the lesson? Return to the Science Probe at the beginning of the lesson. Explain why you agree or disagree with that person now.

EXPLAIN THE PHENOMENON

Revisit your claim about how cells are organized to make up an organism. Review the evidence you collected. Explain how your evidence supports your claim.

START PLANNING

STEM Module Project
Science Challenge

Now that you've learned about the levels of organization within organisms, go back to your Module Project to prepare for the debate. Keep in mind that during the debate you'll want to explain how levels of organization in organisms, such as the glass frog at the beginning of the lesson, form subsystems that work together.

PERFORMANCE EXPECTATION

SCIENCE PROBES

LESSON 2 LAUNCH

Is muscle alive?

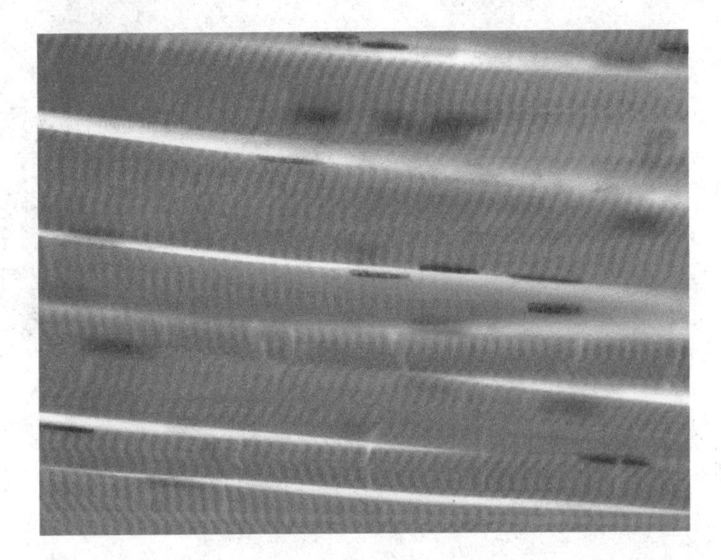

Four students were discussing muscle. They each had a different idea about whether muscles are alive. This is what they said:

Millie: I think muscles are living because they are inside of our bodies.

Akhim: I think muscles are living because they are made up of cells.

Bao: I think muscles are nonliving because their cells do not reproduce.

Tony: I think muscles are nonliving because they do not exist as single-celled organisms.

Circle the student you most agree with. Explain why you agree with that person.

____

____

____

____

____

____

You will revisit your response to the Science Probe at the end of the lesson.

LESSON 2

Structure and Support

ENCOUNTER THE PHENOMENON | How is this girl able to do a handstand?

One part of the body that enables the girl to to a handstand is her muscles. Do you think she controls all of her muscles?

Shake hands with another student. What muscles did you use? Did you have to think about this action? Explain your answer.

Rest your index and middle fingers on the thumb side of your wrist until you can feel your pulse. What muscles control your pulse? Can you change the speed of your pulse by thinking about it?

What parts of your body have muscles that you can control by thinking about them? What are their functions?

Think of other muscles in your body, besides your heart, that work without you thinking about them. How do the functions of these muscles differ from the ones you consciously control?

Dance Moves

GO ONLINE Watch the video *Dance Moves* to see this phenomenon in action.

EXPLAIN THE PHENOMENON

Did you see how your muscles have different functions? Use your observations about the phenomenon to make a claim about what body systems work together to provide structure and support.

CLAIM

Organisms are structured and supported by...

COLLECT EVIDENCE as you work through the lesson.

Then return to these pages to record your evidence.

EVIDENCE

A. What evidence have you discovered to explain how muscles and bones provide structure and support for organisms, such as the girl doing the handstand?

B. What evidence have you discovered to explain some ways that different types of animals are provided with structure and support?

MORE EVIDENCE

C. What evidence have you discovered to explain the systems that plants have that provide them with structure and support?

When you are finished with the lesson, review your evidence. If necessary, based on the evidence, revise your claim.

REVISED CLAIM

Organisms are structured and supported by...

Finally, explain your reasoning for how and why your evidence supports your claim.

REASONING

The evidence I collected supports my claim because...

What supports a body and enables it to move?

You probably have already learned that you need muscles in order to move. At the beginning of the lesson, you saw a girl support the weight of her body with just her arms by using her muscles. How do muscles work?

A **muscle** is made of strong tissue that can contract in an orderly way. When a muscle contracts, the cells of the muscle become shorter. When the muscle relaxes, the cells return to their original length.

Relaxed

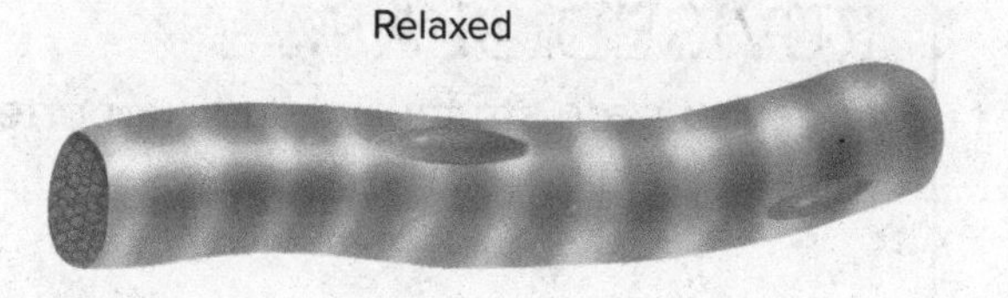

Contracted

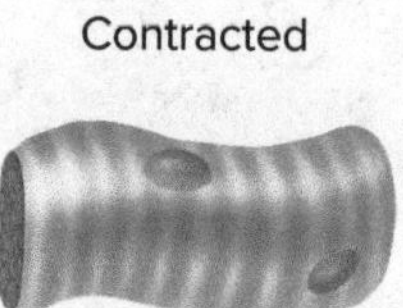

You might recall that mitochondria are the main energy producers in a cell. Because so much energy is required for muscle function, muscle cells are packed with mitochondria.

Muscles allow for movement of the body, but they do not work alone. Brainstorm and write or illustrate what other systems help the body move.

THREE-DIMENSIONAL THINKING

Muscles allow for movement of the body, but they do not work alone. Brainstorm and write or illustrate to **explain** how other **systems** help the body move.

Muscles enable the body to move, but cannot function without the support of bones. Bones can move because they are attached to muscles. The skeletal system and the muscular system work together and move your body.

Want more information?

Go online to read more about body systems that provide structure and support.

FOLDABLES®

Go to the Foldables® library to make a Foldable® that will help you take notes while reading this lesson.

Joints Your bones work together at places called joints. A **joint** is where two or more bones meet. Joints provide flexibility and movement, like you saw with the girl doing a handstand at the start of the lesson. Bones are connected to other bones by tissues called **ligaments.** When the bones in joints move, ligaments stretch and keep the bones from shifting away from each other.

PHYSICAL SCIENCE Connection Your arms and legs may not seem like machines, but in fact they are. When muscles pull on bones they act like a simple machine called a lever. Levers rotate about a fixed point, which in your body are the joints. Three different types of joints enable rotation—ball and socket, hinge, and pivot joints. Examine the different joints in the table.

Types of Movable Joints

Joint	Description	Example
Ligaments Ball and socket	allows bones to move and rotate in nearly all directions	hips and shoulder
Hinge	allows bones to move back and forth in a single direction	fingers, elbows, knees
Pivot	allows bones to rotate	neck, lower arm below the elbow

COLLECT EVIDENCE

How do muscles and bones provide structure and support for organisms, such as the girl doing the handstand at the beginning of the lesson? Record your evidence (A) in the chart at the beginning of the lesson.

The skeletal system does more than help the body move. Learn more about the function of bones in the body.

LAB Make No Bones About It

You have learned that the skeleton, along with muscles, enables movement. But what else does the skeleton do for the body?

Safety  

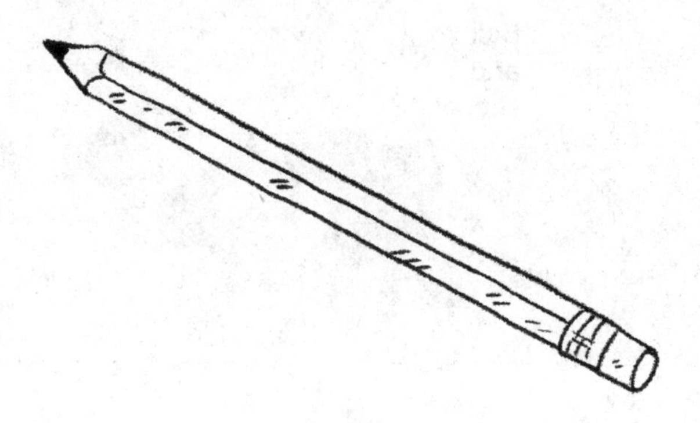

Materials

bubble wrap

plastic bag

plastic jar

Procedure

1. Read and complete a lab safety form.
2. Place one piece of bubble wrap in a plastic bag.
3. Firmly squeeze the plastic bag for five seconds, popping as many bubbles as you can. Remove the bubble wrap and count how many bubbles are popped. Record the data in the Data and Observations section below.
4. Repeat step 3 with the bubble wrap in the plastic jar.
5. Follow your teacher's instructions for proper cleanup.

Data and Observations

Analyze and Conclude

6. Infer which function of the skeletal system this experiment modeled.

Bones provide support. They help you sit up, stand, and raise your arm over your head to ask a question. What else can the skeleton do for the body?

Protection Feel your head, and then feel your stomach. Your stomach is softer than your head. The hard, rigid structure you feel in your head is your skull. It protects the soft, fragile tissue of your brain from damage. Other bones protect the spinal cord, heart, lungs, and other internal organs.

Production and Storage Another function of bones is to produce and store materials needed by your body. Red blood cells are produced inside your bones. Bones store fat and calcium. Calcium is needed for strong bones and for many cellular processes.

THREE-DIMENSIONAL THINKING

Use an **argument** to support or refute the following claim: Your body would be able to **function** without your skeletal **system.**

What about organisms that don't have bones, such as worms? What provides them with structure, and how do they move?

In what ways are different animals supported and provided with structure?

You move by using your muscles and skeleton. However, an earthworm does not have a skeleton. How is an earthworm able to move?

LAB Exploring Earthworm Movement

Safety 

Materials

paper towel
plastic container
earthworm

Procedure

1. Read and complete a lab safety form.
2. Place a paper towel in the bottom of a plastic container. Add water to the paper towel until it is damp, but not dripping wet.
3. Place an earthworm on the surface of the paper towel and observe the earthworm for several minutes.
4. Pay particular attention to what happens when the earthworm moves. Note the changes in the earthworm's body and the motion that enables it to move.
5. Follow your teacher's instructions for proper cleanup.

Analyze and Conclude

6. How do you think the body of an earthworm is provided with structure even though it has no skeleton?

Fluid Support Some animals have a **hydrostatic skeleton,** which is a fluid-filled internal cavity surrounded by muscle tissue. Muscles help the organisms move by pushing the fluid in different directions. Flatworms, such as the one shown to the right, sea anemones (uh NE muh neez), and earthworms are organisms that have hydrostatic skeletons.

External Support Hard outer coverings provide support and protection for many animals. Sometimes called shells, these outer coverings support animals such as crabs, snails, and the scorpion shown to the left. A thick, hard outer covering that protects and supports an animal's body is called an **exoskeleton.**

THREE-DIMENSIONAL THINKING

Compare and contrast the structure and function of hydrostatic skeletons and exoskeletons.

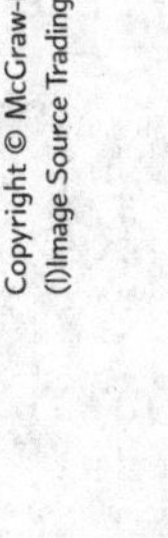

COLLECT EVIDENCE

How are different types of animals provided with structure and support? Record your evidence (B) in the chart at the beginning of the lesson.

As you just learned, some animals are provided with structure and can move without rigid bones or shells. Think about a squid, for example, and how it is provided with structure and is able to move. Learn more on the next page.

HOW NATURE WORKS

Jet Propulsion

The Secret of a Squid's Speed

A squid swims slowly along the ocean floor, flapping its delicate fins. Suddenly, it spots a shark approaching. In a flash, the squid darts away and is gone. When a squid has to move fast, its fins can't get the job done. It uses jet propulsion.

Think of what happens when you let go of a balloon as you're blowing it up. Air rushes out of the balloon in one direction, launching it in the opposite direction. This movement is an example of jet propulsion. Cephalopods (SE fuh luh podz), animals such as squids, cuttlefishes, and octopuses, use jet propulsion to move quickly through the ocean. However, they shoot water out of their bodies instead of air.

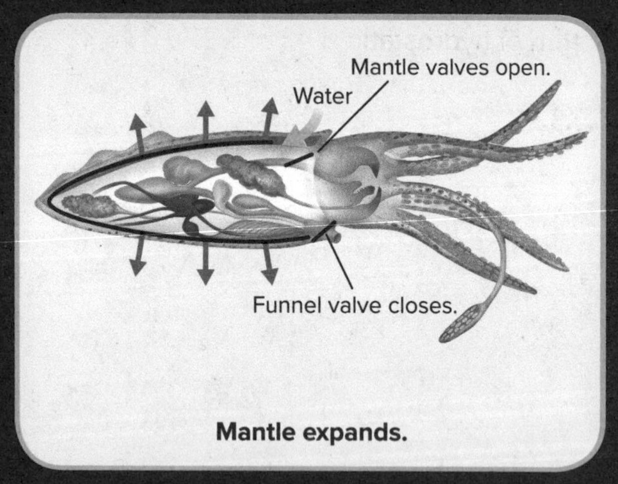

A A very important part of a squid's body is called the mantle, which is the wall that encloses all body organs. When the squid opens its mantle valves, drawing in water, the mantle expands. Then the mantle valves close so the water can't escape.

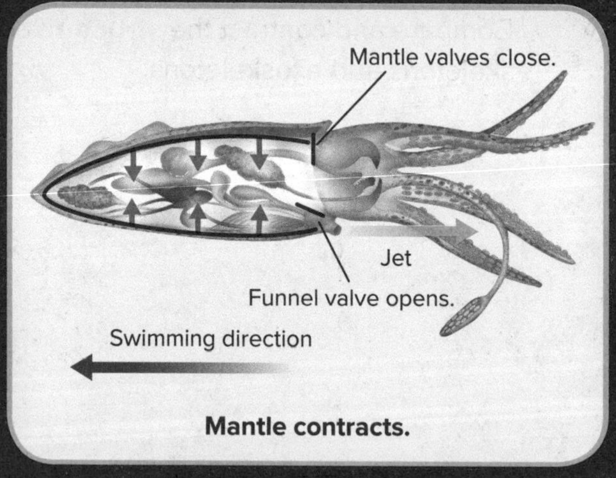

B The squid contracts its mantle, opens its funnel valve, and shoots out water through the funnel. This propels the squid through the water in the opposite direction. A squid can change directions by bending its funnel the other way.

It's Your Turn

Diagram Research another animal that uses jet propulsion. Then draw and label a diagram that explains how it uses jet propulsion to move.

What do different types of muscles do?

You have learned how muscles enable organisms to move. However, muscles serve other functions in the body besides movement. Three types of muscle cells—skeletal, cardiac, and smooth—have different characteristics and functions in the body. In this activity, you will examine cells of each type and observe their similarities and differences.

 Types of Muscles

Safety 

Materials

compound microscope

prepared slide of skeletal muscle

prepared slide of cardiac muscle

prepared slide of smooth muscle

Procedure

1. Read and complete a lab safety form.
2. Place a prepared slide of skeletal muscle cells on the stage of a compound light microscope.
3. In the Data and Observations table on the next page, describe the cells' shape and color. Note any differences among them.
4. Locate the nucleus in one of the cells. Where is it in the cell? Do any cells have more than one nucleus? Record your observations in the table.
5. Describe the arrangement of the cells in the tissue. Do they have a pattern, or are they randomly arranged? Record your observations in the table.
6. Repeat steps 3 through 6 for the prepared slides of cardiac muscle and smooth muscle.
7. Follow your teacher's instructions for proper cleanup.

Data and Observations

Observations of Three Types of Muscle Cells			
Cell Characteristics	Skeletal	Cardiac	Smooth
Shape and color			
Nucleus location and number			
Arrangement patterns			

Analyze and Conclude

8. How do the characteristics of each cell type differ?

9. Based on your observations of the structure of each type of cell, can you infer any of their functions?

You just examined three types of muscle cells. Read about each type to learn more!

Skeletal Muscle The type of muscle that attaches to bones is skeletal muscle. Skeletal muscles are also called voluntary muscles, which are muscles that you can consciously control. The contractions of skeletal muscles can be quick and powerful, such as when you run fast.

Cardiac Muscle Your heart is made of **cardiac muscles,** which are found only in the heart. A cardiac muscle is a type of involuntary muscle, which is muscle you cannot consciously control. When cardiac muscles contract and relax, they pump blood through your heart and through blood vessels throughout your body.

Smooth Muscle Blood vessels and many organs, such as the stomach, are lined with smooth muscles. **Smooth muscles** are involuntary muscles named for their smooth appearance.

What systems do plants have that give them structure?

Many animals use muscles to move while bones offer support and structure. Do plants have similar structures that give them support?

INVESTIGATION

Plant Posture

Take a look at the images of plant structures below. Use what you know about animal bodies to infer how the indicated structures support the plant.

Roots Even though the roots of most plants are never seen, they are vital to a plant's survival. Roots anchor a plant, either in soil or onto another plant or an object such as a rock. All roots help a plant stay upright. Some plants have roots that spread out in all directions several meters from a plant's stem. All root systems help a plant absorb water and other substances from the soil.

Many plants have a large main root, called a taproot, with smaller roots growing from it. Some plants have additional small roots above ground, called prop roots, that help support the plant. Other plants have fibrous root systems that consist of many small branching roots.

Taproot

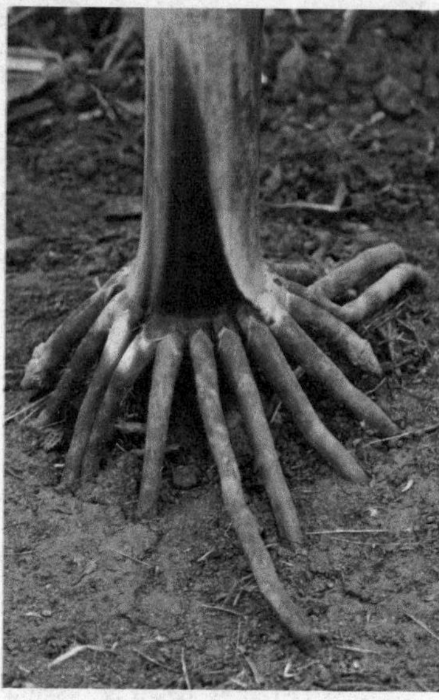
Prop Roots

Fibrous Roots

Plants such as radishes and carrots store food in their roots. This food can be used to grow new plant tissues after a dry period or a cold season. Sugar stored in the roots of sugar maple trees over the winter is converted to maple sap in the spring. Farmers drain some of the sap from these trees and boil it to make maple syrup.

INVESTIGATION

Putting Down Roots

Research a plant in your local environment and determine the type of root system it has. Write about how its root system benefits the whole plant.

Stems Have you ever leaned against a tree? If so, you were leaning on a plant stem. In plants such as the tree, the stem is obvious. Other plants, such as the potato and the iris, have underground stems that are often mistaken for roots.

Stems support branches and leaves. Their tissues transport water, minerals, and food. The sugar produced during photosynthesis flows through the stem to all parts of a plant. Another important function of stems is the production of new cells for growth, but only certain regions of a stem produce new cells.

Plant stems usually are classified as either herbaceous or woody. Woody stems are stiff and typically not green. Trees and shrubs have woody stems. Herbaceous stems usually are soft and green.

THREE-DIMENSIONAL THINKING

Examine the image and determine which plant has a woody stem structure and which has an herbaceous stem structure. **Explain** your reasoning.

COLLECT EVIDENCE

How do systems provide plants with structure and support? Record your evidence (C) in the chart at the beginning of the lesson.

THREE-DIMENSIONAL THINKING

Use the space below to compare and contrast the **systems** that provide **structure** and support in plants and animals.

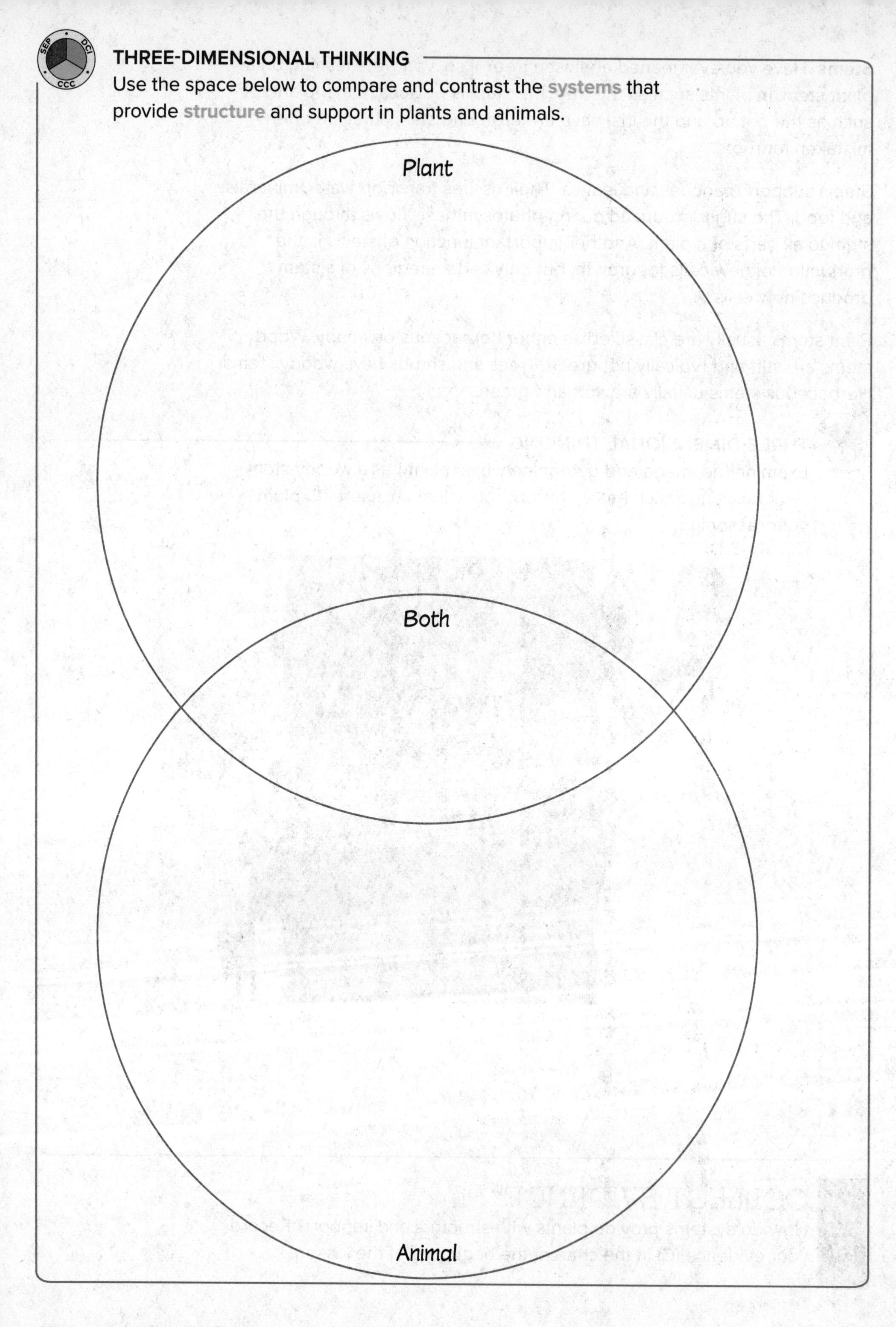

HOW IT WORKS

A Bionic Arm

How brains control mechanical arms

Imagine what your arms would be like without muscles. You would not be able to control them. For many years prosthetic, or artificial, arms looked real, but they didn't work like real arms. Recently scientists have developed a bionic, or mechanical, arm. Signals from the patient's brain control it.

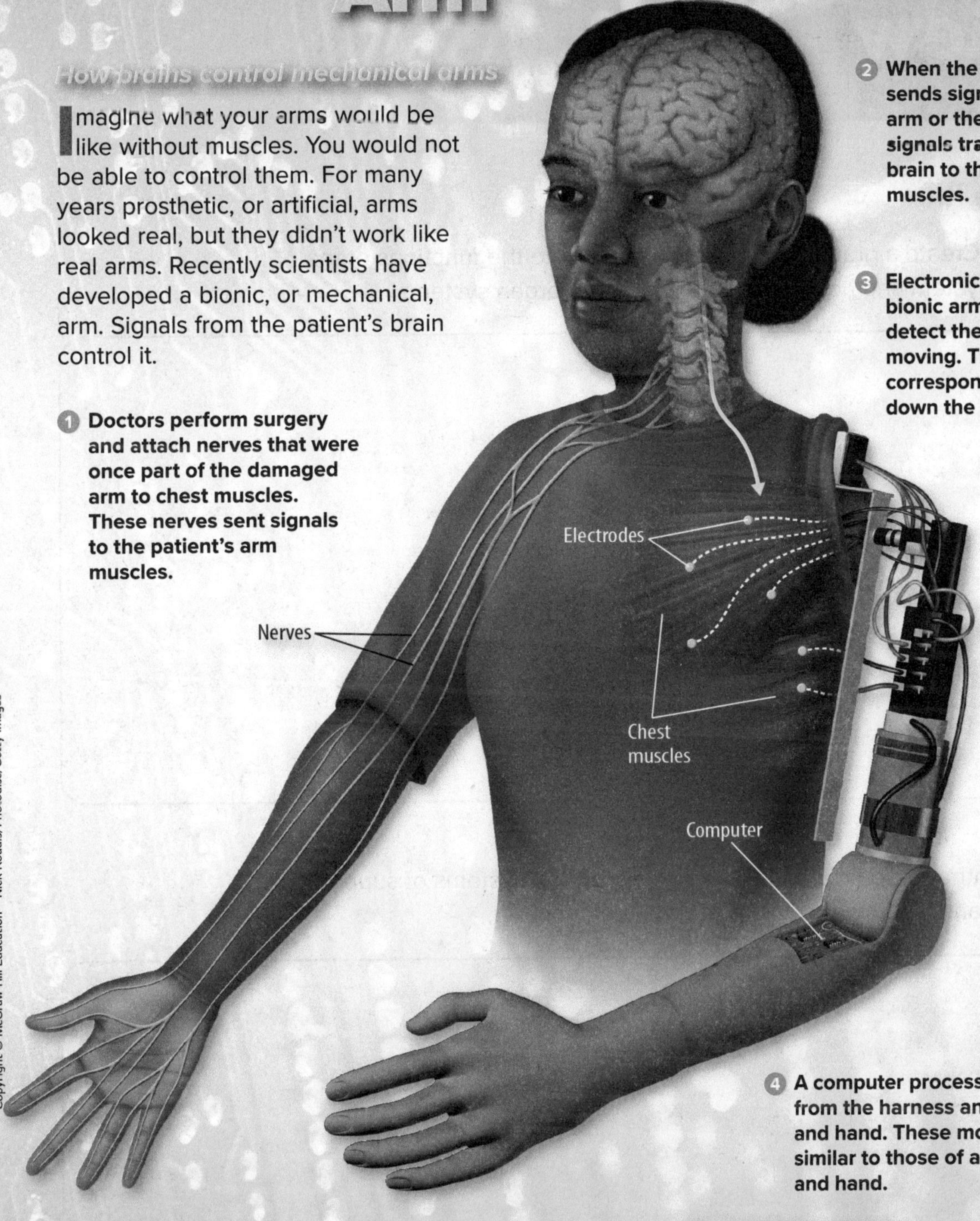

1. **Doctors perform surgery and attach nerves that were once part of the damaged arm to chest muscles. These nerves sent signals to the patient's arm muscles.**
2. **When the patient's brain sends signals to move the arm or the hand, the signals travel from the brain to the chest muscles.**
3. **Electronic sensors in the bionic arm's harness detect the chest muscle moving. The sensors send corresponding signals down the bionic arm.**
4. **A computer processes the signals from the harness and moves the arm and hand. These movements are similar to those of a biological arm and hand.**

Electrodes
Nerves
Chest muscles
Computer

It's Your Turn

Collaborate With a partner, discuss how being open to new ideas allows scientists and engineers to create useful technologies for people. Summarize your ideas in a paragraph.

LESSON 2

Review

Summarize It!

1. **Organize** Create a graphic organizer to summarize the functions of the muscular system and how it interacts with other organ systems.

2. **Create** another graphic organizer to summarize the systems of support found in plants and how they interact with one another.

Three-Dimensional Thinking

3. Why do muscle cells have so many mitochondria?

A Muscle cells are bigger than every other cell and can fit more mitochondria.

B Muscle cells need to quickly respond to energy needs.

C The mitochondria in muscle cells are smaller so more are needed.

D There are more mitochondria only because there are more nuclei.

---

4. What is the effect when a muscle contracts?

A The muscle lengthens.

B The muscle pushes on a bone.

C The muscle pushes on another muscle.

D The muscle shortens.

Use the diagram below to answer question 5.

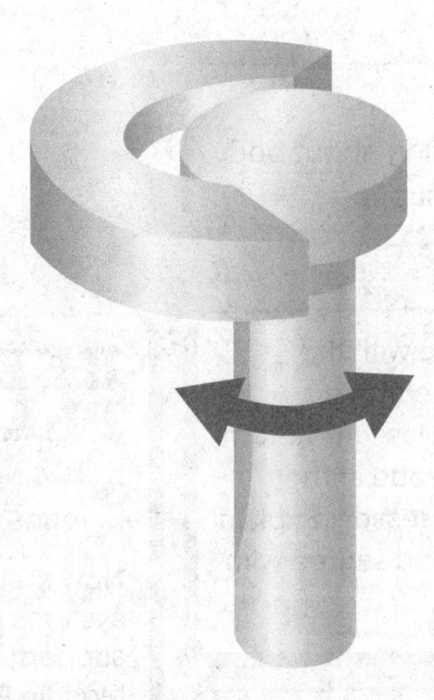

5. The image above represents a joint that would be found in which structure?

A finger

B knee

C neck

D shoulder

Real-World Connection

6. **Infer** Sometimes people who have issues with their joints require joint replacement surgery. The joint is replaced with an artificial joint. Not all artificial joints are the same. Infer why there are different types of artificial joints.

7. **Argue** A city sidewalk inspector is considering installing rubber sidewalks in his city after noticing how concrete sidewalks cracked near the base of trees. Construct an argument in support of this plan, explaining why rubber would be an improvement.

Still have questions?
Go online to check your understanding about body systems that provide structure and support.

REVISIT SCIENCE PROBES

Do you still agree with the person you chose at the beginning of the lesson? Return to the Science Probe at the beginning of the lesson. Explain why you agree or disagree with that person now.

EXPLAIN THE PHENOMENON

Revisit your claim about how the body is structured and supported. Review the evidence you collected. Explain how your evidence supports your claim.

KEEP PLANNING

STEM Module Project
Science Challenge

Now that you've learned about systems of structure and support, go back to your Module Project to keep planning. Explain in your debate how systems of structure and support in organisms, such as the glass frog, interact with other body systems.

LESSON 3 LAUNCH

PAGE KEELEY SCIENCE PROBES

Digestion and Food

The cells in our body need a source of energy to carry out their cell functions. They also need building blocks for growth and repair of tissues. The energy and building blocks come from food digested by the digestive system. Put an X next to all the things that our cells get from the digestive system to use for energy and building blocks.

_________ water

_________ molecules of sugar

_________ bread

_________ vitamins

_________ calcium

_________ molecules of protein

_________ diet soda

_________ banana

_________ carbon dioxide

_________ hamburger

_________ molecules of fat

_________ carrots

_________ rice

Explain your thinking. What rule or reasoning did you use to decide what cells use for energy and building blocks?

____

____

____

____

____

You will revisit your response to the Science Probe at the end of the lesson.

LESSON 3

Obtaining Energy and Removing Waste

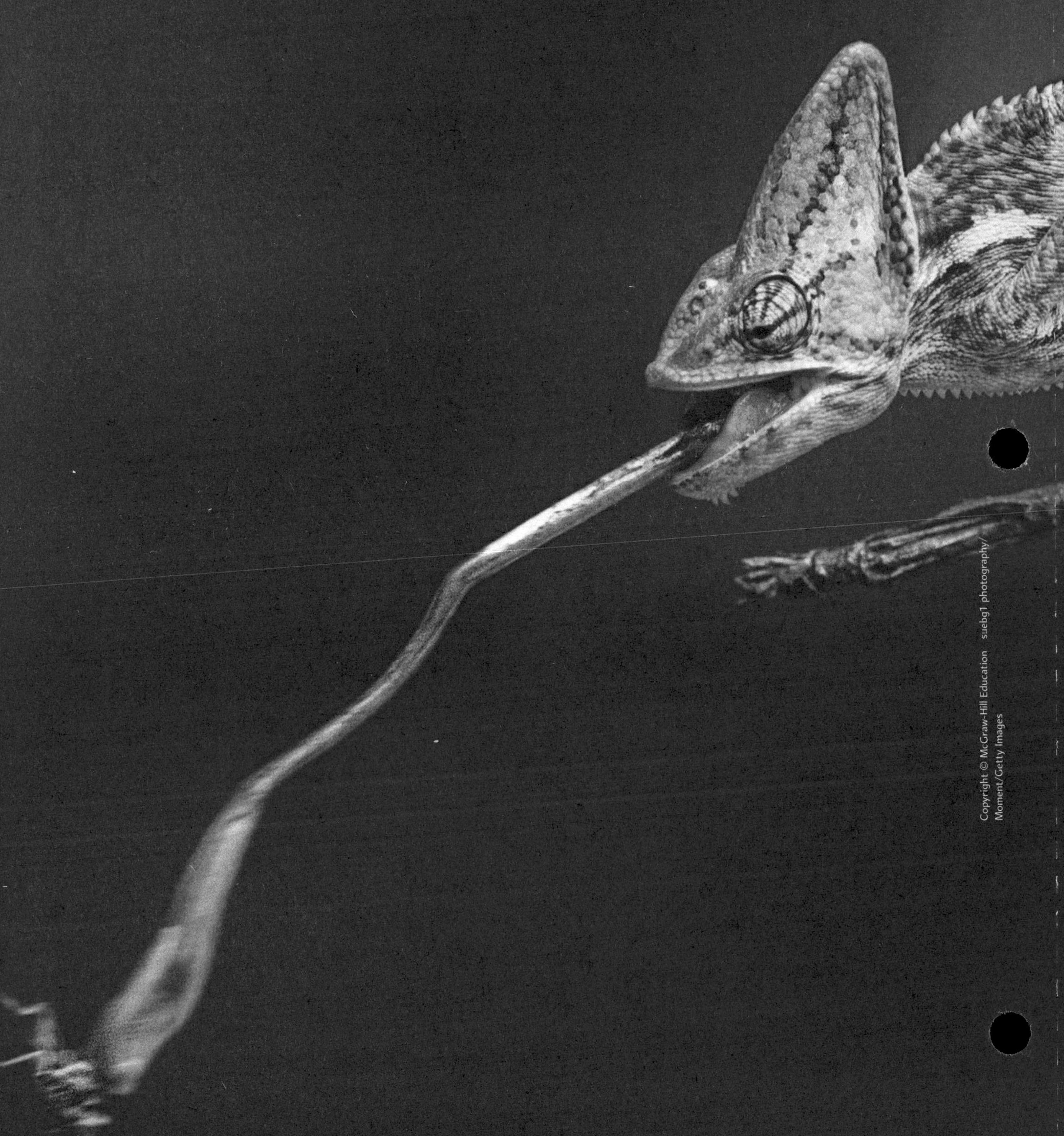

ENCOUNTER THE PHENOMENON | What happens to this insect after it is eaten?

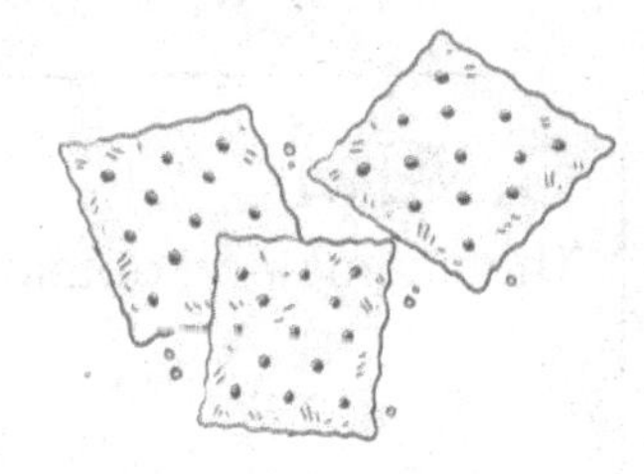

Like the chameleon eating the insect, humans also eat to obtain energy and survive. Discover more about the process of obtaining energy.

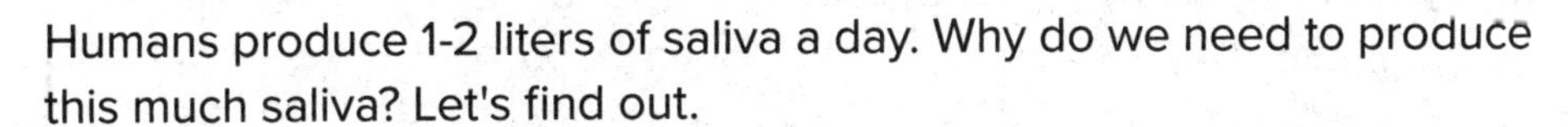

Humans produce 1-2 liters of saliva a day. Why do we need to produce this much saliva? Let's find out.

Eat a cracker. Record your observations.

Put a cracker in your mouth for a minute and a half, and then eat it. Record your observations. How was eating this cracker different than eating the first cracker?

What role do you think saliva plays in digestion?

Insect Lunch

GO ONLINE Watch the video *Insect Lunch* to see this phenomenon in action.

EXPLAIN
THE PHENOMENON

Did you see how your saliva helped break down the cracker? Use your observations about the phenomenon to make a claim about the process of breaking down food to fuel the body's activities.

CLAIM

The chameleon obtains energy from food and gets rid of wastes through...

COLLECT EVIDENCE as you work through the lesson.

Then return to these pages to record your evidence.

EVIDENCE

A. What evidence have you discovered to explain why digestion is important for organisms, such as the chameleon?

B. What evidence have you discovered to explain what happens after food is digested and the nutrients are obtained?

MORE EVIDENCE

C. What evidence have you discovered to explain how plants obtain energy?

When you are finished with the lesson, review your evidence. If necessary, based on the evidence, revise your claim.

REVISED CLAIM

The chameleon obtains energy from food and gets rid of wastes through...

Finally, explain your reasoning for how and why your evidence supports your claim.

REASONING

The evidence I collected supports my claim because...

Why do organisms eat?

How do you decide what to eat or when to eat? Although you can survive for weeks without food, you might become hungry within hours of your last meal. Hunger is your body's way of telling you that it needs food. Why does your body need food?

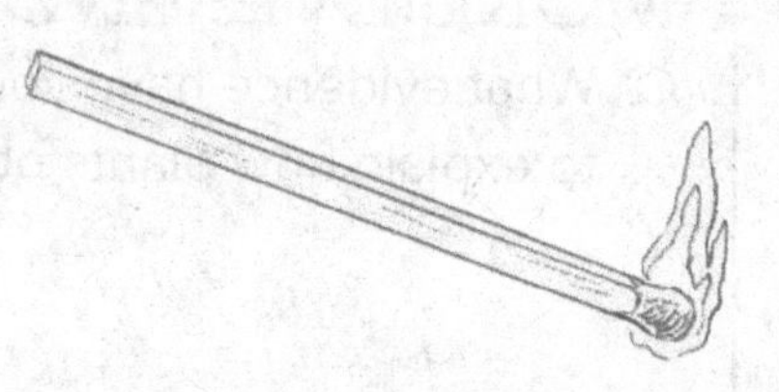

INVESTIGATION

Learning S'more About Science

Observe your teacher's demonstration.

1. What happened to the water? Why do you think this happened?

2. What do you think happens to your body when you eat a marshmallow? Why?

Energy from Food The amount of energy in food is measured in Calories. A **Calorie** (Cal) is the amount of energy it takes to raise the temperature of 1 kg of water by 1°C. How much energy do foods contain? Each food is different. One grape contains 2 Cal, but a slice of cheese pizza has 220 Cal. All foods give your body energy to use.

Want more information?
Go online to read more about body systems that enable organisms to obtain energy and remove waste.

FOLDABLES®
Go to the Foldables® library to make a Foldable® that will help you take notes while reading this lesson.

What does energy from food power?

Every activity you do, such as riding a bike or even sleeping, requires energy. Your digestive system processes food and releases energy that is used for cellular processes and all the activities that you do.

The amount of energy a person needs depends on several factors, such as weight, age, activity level, and gender. Playing soccer, for example, requires more energy than playing a video game. How does the food you eat supply you with energy? The energy comes from nutrients.

INVESTIGATION

Using Energy

List some activities that use energy. Describe the activity and explain what systems are involved. Then, rank the amount of energy needed for the activities as 1, 2, or 3, with 1 being the activity that requires the most energy.

Activity	Description	Energy Required

What nutrients are in food?

Food provides your body with nutrients and Calories. **Nutrients** are the parts of food used by the body to grow and survive. Each nutrient is important and has its own function in the body.

INVESTIGATION

You Are What You Eat

1. Using the materials provided by your teacher, search for foods that contain a high amount of your assigned nutrient.
2. Your teacher will ask you to find a number of food items containing your nutrient. Find these items.
3. Once you have found the appropriate number of items, form a group with other students who were assigned the same nutrient.
4. As a group, make a chart listing your food items. Show the amount of your assigned nutrient present in each item. Share your chart with the class.
5. Research and explain the function your assigned nutrient has in the body.

6. **WRITING Connection** Your classmate does not think that it is important to get your nutrient into your diet. Cite specific evidence to support the argument that the body requires your nutrient.

How does a body get nutrients from food?

Nutrients stored in food are necessary for the body to function properly. You may have discovered in the Investigation *You Are What You Eat* that nutrients like protein build muscles while calcium helps strengthen bones. How do these nutrients enter and get processed by the body?

LAB Greatest Thing Since Sliced Bread

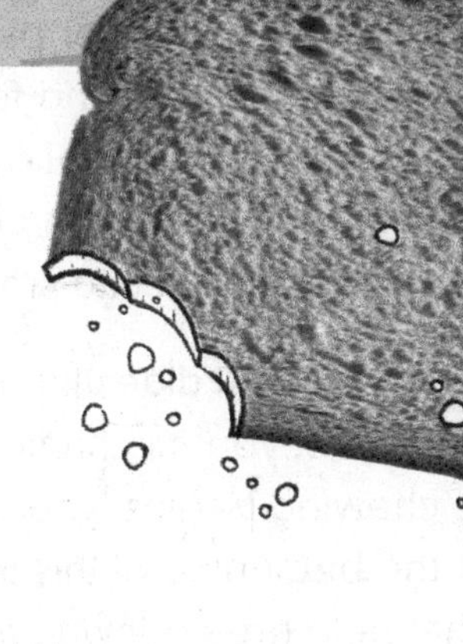

Safety 

Materials

plastic bowl
sliced bread
vinegar
water
towel

Procedure

1. Read and complete a lab safety form.
2. Tear a slice of bread into small pieces.
3. Place the pieces of bread into a bowl and add 1 tablespoon of vinegar and 3 tablespoons of water. With your hands, mash the bread and vinegar mixture.
4. Place the bread mass onto an absorbent towel. Squeeze the bread mixture with the towel over the bowl. Observe the remaining bread.
5. Follow your teacher's instructions for proper cleanup.

Analyze and Conclude

6. What do you think steps 2, 3, and 4 of the procedure represent? Explain your reasoning.

How does digestion work?

By breaking down a slice of bread, you just modeled digestion. **Digestion** is the mechanical and chemical breakdown of food into small particles and molecules that your body can absorb and use. In order for a body to obtain nutrients, food must be digested. Let's suppose that you have eaten the slice of bread from the previous page. We can observe the process of digestion by following the path of the bread through your digestive system.

Types of Digestion Before your body can absorb nutrients from the bread, it must be broken down into small molecules by digestion. There are two types of digestion—mechanical and chemical. In **mechanical digestion,** food is physically broken into smaller pieces. Mechanical digestion happens when you chew, mash, and grind food with your teeth and tongue. Smaller pieces of food are easier to swallow and have more surface area, which helps with chemical digestion. In **chemical digestion,** chemical reactions break down pieces of food into small molecules.

The Mouth Mechanical digestion of food begins in your mouth. Using muscles in your jaw, you mechanically digest your bread as you chew. But even before chewing begins, your mouth prepares for digestion. As you observed at the beginning of this lesson with crackers, saliva contains chemicals that help break down carbohydrates. Saliva also contains substances that neutralize acidic foods.

The Esophagus After you swallow a bite of your bread, it enters your esophagus (ih SAH fuh gus). The **esophagus** is a muscular tube that connects the mouth to the stomach. Food moves through the esophagus and the rest of the digestive tract by waves of muscle contractions, called **peristalsis** (per uh STAHL sus).

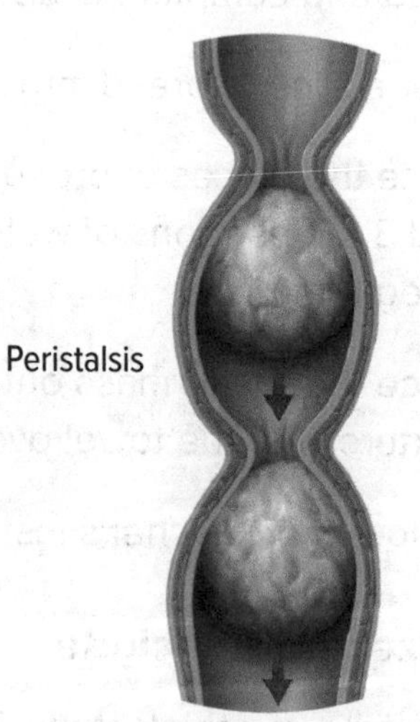
Peristalsis

Peristalsis is similar to squeezing a tube of toothpaste. When you squeeze the bottom of the tube, toothpaste is forced toward the top of the tube. As muscles in the esophagus contract and relax, partially digested food is pushed down the esophagus and into the stomach.

In the Lab *The Greatest Thing Since Sliced Bread* you modeled mechanical digestion when you tore the bread into small pieces. After the bread was physically broken down, the vinegar and water mixture further broke the bread down, much like chemical digestion.

COLLECT EVIDENCE

Why is digestion important for organisms such as the chameleon? Record your evidence (A) in the chart at the beginning of the lesson.

The Stomach Once your partially digested bread leaves the esophagus, it enters the stomach. The stomach is a large, hollow organ. One function of the stomach is to temporarily store food. Another function of the stomach is to aid in chemical digestion. The stomach contains an acidic fluid called gastric juice. Acid helps break down some of the structures that hold plant and animal cells together, including the bread.

Tongue
Salivary glands
Esophagus
Stomach
Small intestine
Large intestine

The Small Intestine Most chemical digestion occurs in the small intestine, a long tube connected to the stomach. It is also where nutrient absorption occurs. The folds of the small intestine are covered with fingerlike projections called **villi** (VIH li) (singular, villus). Each villus contains small blood vessels. Nutrients in the small intestine enter the blood through these blood vessels.

The Large Intestine Most of the water in ingested foods and liquids is absorbed in the small intestine. As the bread travels through the large intestine, even more water is absorbed. Materials that pass through the large intestine are the waste products of digestion. The waste products become more solid as excess water is absorbed. Then the semisolid waste is eliminated from the body.

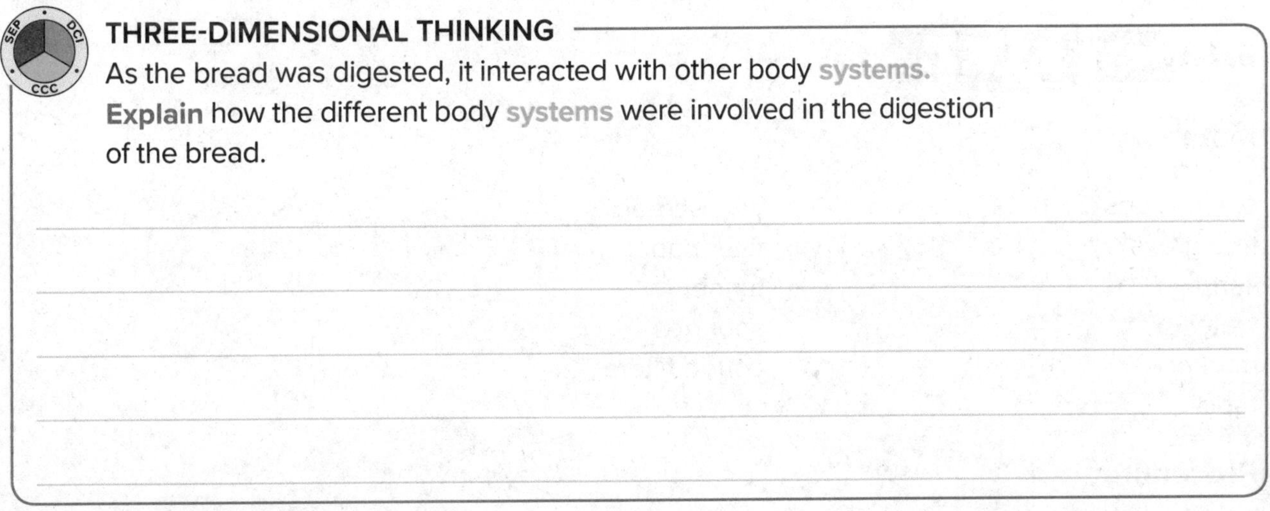

THREE-DIMENSIONAL THINKING

As the bread was digested, it interacted with other body systems. **Explain** how the different body systems were involved in the digestion of the bread.

GO ONLINE for additional opportunities to explore!

Investigate digestion further by performing one of the following activities.

- [ ] **Model** the human digestive system in the **Lab** *Model Digestion from Start to Finish.*

OR

- [ ] **Model** and **investigate** how a gizzard works in the **Lab** *How do gizzards help birds eat?*

Removing Waste When you modeled digestion, you may have noticed that there was leftover bread, representing food waste. The body produces other waste products as well. For example, some water and vinegar remained in the towel, representing liquid waste. Liquid waste is handled by something called the urinary system.

Your body excretes, or eliminates, different substances from different body systems. They are processed by the excretory system. The **excretory system** collects and eliminates wastes from the body and regulates the level of fluid in the body. The excretory system is made of different body systems. For example, you just read that liquid waste is handled by the urinary system. The respiratory system, which you will learn about in the next lesson, releases waste as carbon dioxide. Your skin also removes waste in the form of excess salt and water through sweat glands.

LAB Filtering Waste

You have two kidneys, one on each side of your body. The kidney is an important organ in the excretory system. Model the function of the kidneys to see how they work!

Safety 

Materials

plastic cups (3)
fine gravel
sand
water
marker
tape
wire screen
filter paper
funnel
sponges
coffee filters

Procedure

1. Read and complete a lab safety form.
2. Label three plastic cups 1, 2, 3.
3. Mix a small amount of fine gravel and sand with water in cup 1.
4. Place a small piece of wire screen in a funnel, and place the funnel in cup 2.

5. Carefully pour the sand-water-gravel mixture into the funnel. Let it drain. Record your observations in your Science Notebook.

6. Remove the screen. Replace it with a piece of filter paper. Place the funnel in cup 3.

7. Carefully pour the contents of cup 2 into the funnel. Let it drain. Record your observations.

8. Follow your teacher's instructions for proper cleanup.

Analyze and Conclude

9. Describe what happened during each filtration.

10. Why is the function of the kidney so important for the entire body?

Kidneys The **kidneys** are bean-shaped organs that filter, or remove, waste from blood. They enable harmful substances to be removed from the body. If waste was allowed to build up in the blood, it would become toxic and harm the entire body.

COLLECT EVIDENCE

What happens after food is broken down and nutrients are absorbed? Record your evidence (B) in the chart at the beginning of the lesson.

A Closer Look: Celiac Disease

While at the grocery store, have you ever noticed that some foods are marked gluten-free? Gluten is a protein that is found in wheat, rye, and barley. It may also be found in medicines, adhesives on envelopes, and lip balm. Individuals with celiac disease cannot eat gluten because it will cause damage to their small intestines.

Celiac disease can affect each person differently. Some individuals have symptoms such as abdominal pain, mouth ulcers and distension, which is the expansion of the stomach and waist caused by gas or fluids. Other individuals may have no symptoms at all.

Changes in the digestive tract make it difficult for the body to absorb necessary nutrients. This is called malabsorption. Malabsorption can cause weight loss and fatigue because the intestine cannot absorb carbohydrates and fats. Anemia may also develop from a deficiency in iron. Additionally, calcium and vitamin D malabsorption may be responsible for osteoporosis, which is a weakening in the bones.

Treatment for celiac disease is a diet free of gluten. Many stores offer gluten-free options for nutritious foods like bread and cereal.

It's Your Turn

READING Connection Research claims about gluten and Celiac disease. Choose one text that makes a claim based on reasons and evidence and one that does not. Compare the scientific accuracy of the texts in your Science Notebook.

How do plant bodies obtain energy and get rid of waste?

Food, water, and oxygen are three things you need to survive. Some of your organ systems process these materials, and others transport them throughout your body. Like you, plants need food, water, and oxygen to survive. Unlike you, plants do not take in food. Most of them make their own.

Leaves are the major food-producing organs of plants. This means that leaves are the site of photosynthesis (foh toh SIHN thuh sus). **Photosynthesis** is a series of chemical reactions that convert light energy, water, and carbon dioxide into the food-energy molecule glucose and give off oxygen. Glucose moves out of food-making cells, enters a tissue called phloem, and flows to all plant cells. Cells then break down the sugar and release energy.

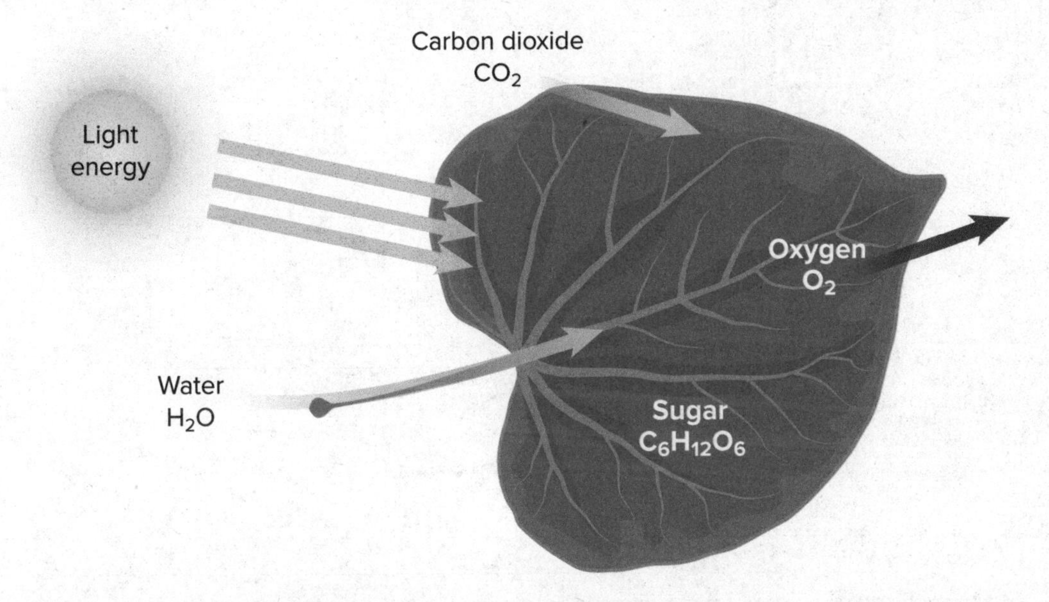

Like animal bodies, plants also require water to survive. After water enters a plant's roots, it moves into a tissue called xylem. Water then flows inside xylem to all parts of a plant. Like you, plants produce water vapor as a waste product. Carbon dioxide, oxygen, and water vapor pass into and out of a plant through tiny openings in leaves.

THREE-DIMENSIONAL THINKING

In your Science Notebook write an **argument** supported by evidence to support or refute the following claim: I obtain energy in the same way that a plant obtains **energy.**

COLLECT EVIDENCE

How do plants obtain energy? Record your evidence (C) in the chart at the beginning of the lesson.

LESSON 3

Review

Summarize It!

1. **Write** a short speech to present to an elementary class explaining to them the importance of proper nutrition. Provide information about why nutrition is important for the whole body, including ways that nutrients are transported.

Three-Dimensional Thinking

2. Peristalsis occurs in the esophagus and helps food travel from the mouth to the stomach. Which of the following explains the cause of peristalsis?

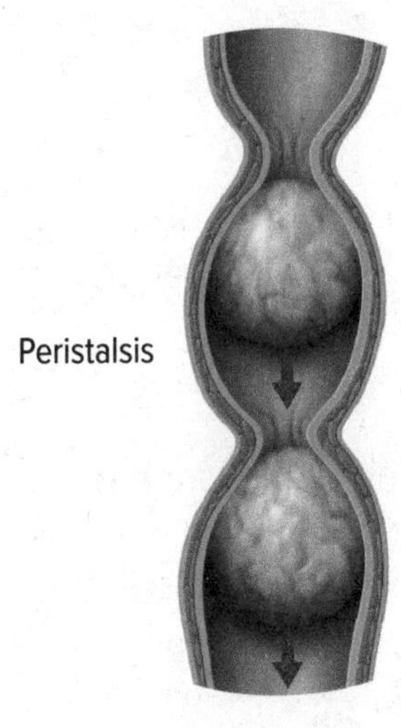

A The skeletal system presses into the esophagus, moving food downward.

B When the lungs inhale and exhale, it can force the esophagus to move.

C Peristalsis is the persistant movement of the esophagus, it is not caused or trigged by any one body system.

D Waves of muscle contractions help push food down the esophagus toward the stomach.

---

3. What would be the effect on your body if you did not have kidneys, important organs in the excretory system?

A My body would not absorb nutrients.

B Waste would collect in the blood and become toxic.

C I could not eat gluten.

D Saliva could not be produced.

Real-World Connection

4. **Construct an Argument** Your friend thinks that eating healthy and balanced meals are not necessary. Construct an argument detailing the importance of obtaining a variety of nutrients and how this can affect your friend's life.

5. **Explain** When someone does not have functioning kidneys, he goes through dialysis, which is a form of treatment that artificially filters his blood through a machine. Explain why such a treatment is necessary.

Still have questions?

Go online to check you understanding about body systems that enable organisms to obtain energy and remove waste.

REVISIT

Do you still agree with the choices you selected at the beginning of the lesson? Return to the Science Probe at the beginning of the lesson. Explain why you agree or disagree with those choices now.

EXPLAIN THE PHENOMENON

Revisit your claim about what happens to food once it is eaten. Review the evidence you collected. Explain how your evidence supports your claim.

KEEP PLANNING

STEM Module Project Science Challenge

Now that you've learned about how organisms obtain energy and remove waste, go back to your Module Project to continue preparing for your debate. You want to explain how these systems in organisms, such as the glass frog, interact with other body systems.

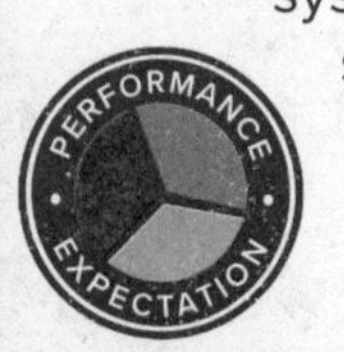

PAGE KEELEY SCIENCE PROBES

LESSON 4 LAUNCH

Moving Blood

Three friends were arguing about the circulatory system. They had different ideas about how blood gets transported to different parts of an animal's body. This is what they said:

Melva: I think all animals need blood vessels to move blood from one part of the body to another.

Zach: I think some animals need blood vessels to move blood from one part of the body to another.

Dennis: I think no animals need blood vessels to move blood from one part of the body to another.

Which friend do you agree with the most? _______________ Explain why you agree with that friend.

____

____

____

____

____

____

You will revisit your response to the Science Probe at the end of the lesson.

LESSON 4

Moving Materials

ENCOUNTER THE PHENOMENON | How does the dye move from the water to the flower petals?

Observe the demonstration performed by your teacher. Use the space below to brainstorm what you think is happening and why.

Do you think a similar process occurs in animals? How?

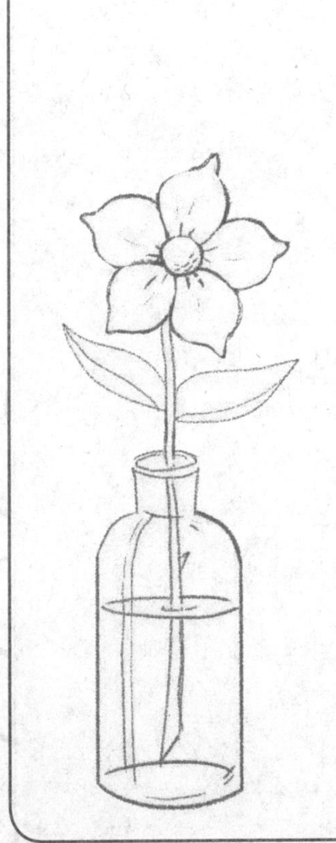

Color-Changing Flowers

GO ONLINE Watch the video *Color-Changing Flowers* to see the phenomenon in action.

EXPLAIN THE PHENOMENON

Did you see how the flower changed colors? Use your observations about the phenomenon to make a claim about how materials are transported through body systems.

CLAIM

Materials are transported through body systems by...

COLLECT EVIDENCE as you work through the lesson. Then return to these pages to record your evidence.

EVIDENCE

A. What evidence have you discovered to explain how plants, such as the flowers on the previous page, transport materials?

B. What evidence have you discovered to explain how humans transport materials?

MORE EVIDENCE

C. What evidence have you discovered to explain how other animals transport materials?

When you are finished with the lesson, review your evidence. If necessary, based on the evidence, revise your claim.

REVISED CLAIM

Materials are transported through body systems by...

Finally, explain your reasoning for how and why your evidence supports your claim.

REASONING

The evidence I collected supports my claim because...

How do plants transport materials?

The flower you observed transported dyed water from the glass to its leaves. How do plants transport the materials they need from their environment through their bodies?

INVESTIGATION

Turning Over a New Leaf

Collect a fresh leaf and place it in a bowl of water. Wait 30 minutes, then observe the leaf. Describe or illustrate your observations below. Why do you think this is happening?

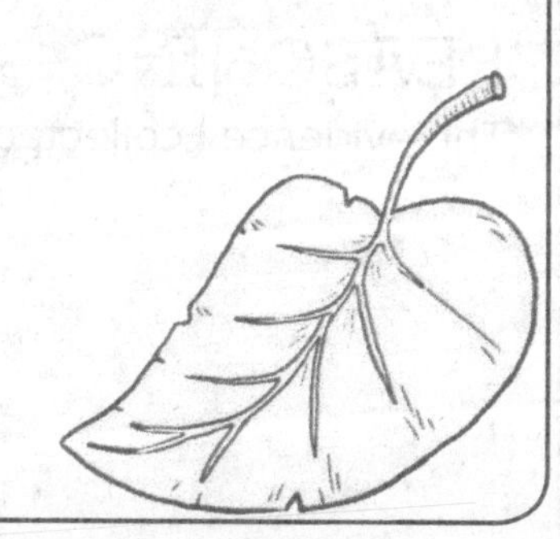

Want more information?
Go online to read more about body systems that transport materials.

FOLDABLES
Go to the Foldables® library to make a Foldable® that will help you take notes while reading this lesson.

Moving Materials Inside Plants You just observed parts of plants transporting materials. How does an entire plant work together as a system to transport materials?

For a plant to survive, water and nutrients must move throughout its tissues. In some plants, these materials can move from cell to cell by the processes of osmosis and diffusion. This means that water and other materials move from areas of high concentration to areas of low concentration. However, most plants such as grasses and trees have specialized tissues in their stems called vascular tissue. **Vascular tissue** is specialized plant tissue composed of tubelike cells that transport water and nutrients in some plants.

Vascular tissue is found in the roots and stems of trees.

One type of vascular tissue—**xylem** (ZI lum)—carries water and dissolved nutrients from the roots to the stem and the leaves. Due to the thickened cell walls of some xylem cells, this tissue also provides support for a plant. Another type of vascular tissue—**phloem** (FLOH em)— carries dissolved sugars throughout a plant.

After water enters a plant's roots, it moves into xylem. Water then flows inside xylem to all parts of a plant. Most plants make their own food—a liquid sugar. The liquid sugar moves out of food-making cells, enters phloem, and flows to all plant cells.

Carbon dioxide, oxygen, and water vapor pass into and out of a plant through small openings in the epidermis, or surface layer, of a leaf called **stomata** (STOH ma tah; singular, stoma). This is shown in the figure to the right. Plants require oxygen and carbon dioxide to make food. Like you, plants produce water vapor as a waste product. This process is called transpiration.

Bubbles formed on the leaf in the investigation on the previous page because gases passed through its stomata.

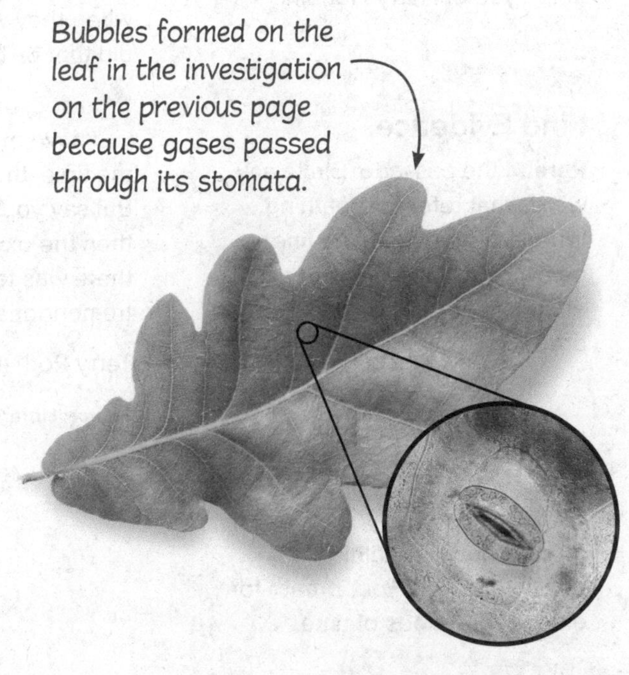

LM Magnification: Unavailable

COLLECT EVIDENCE

How do plants, such as the flowers at the beginning of the lesson, transport materials? Record your evidence (A) in the chart at the beginning of the lesson.

How do humans transport materials?

Breathing is the movement of air into and out of the lungs. Breathing enables your respiratory system to take in oxygen and to eliminate carbon dioxide.

Read a Scientific Text

HISTORY Connection Harry Houdini was a world renowned magician and escape artist who lived from 1874–1926. Houdini routinely performed daring escapes that involved being handcuffed, locked in a container, and submerged in water. How was he able to survive long enough to escape?

CLOSE READING

Inspect

Read the interview with a former employee of Harry Houdini.

Find Evidence

Reread the passage. Circle any words that refer to breathing. Write a sentence on the line below explaining the cause and effect of controlling your breathing.

Make Connections

Collaborate With your partner, discuss how you think it is possible to hold your breath for extended periods of time.

PRIMARY SOURCE

Folklore of Stage People

"Maybe you were too young, but do you remember [Houdini's] famous burying himself alive trick? He used to have himself buried under sand, or in a glass coffin in a tank of water, and when they would pull him up 35 minutes later he was alive. He did that by breath control."

"You see, there's a [certain] amount of oxygen in that coffin. If you have breath control and breathe very slowly and calm all the time, there will be enough oxygen there for 35 minutes. But say you're excited and you start [gasping] for breath, why then the oxygen is used up in a [couple of] minutes. That's all there was to it. But it takes a man of iron nerves and a tremendous amount of control to be able to do it"

Terry Roth (Interviewer) & George Nagle (Interviewee), 1939

Source: Library of Congress

The Respiratory System Gases are exchanged between the body and the environment. The parts of the respiratory system, shown to the right, work together and supply the body with oxygen. They also rid the body of wastes, such as carbon dioxide.

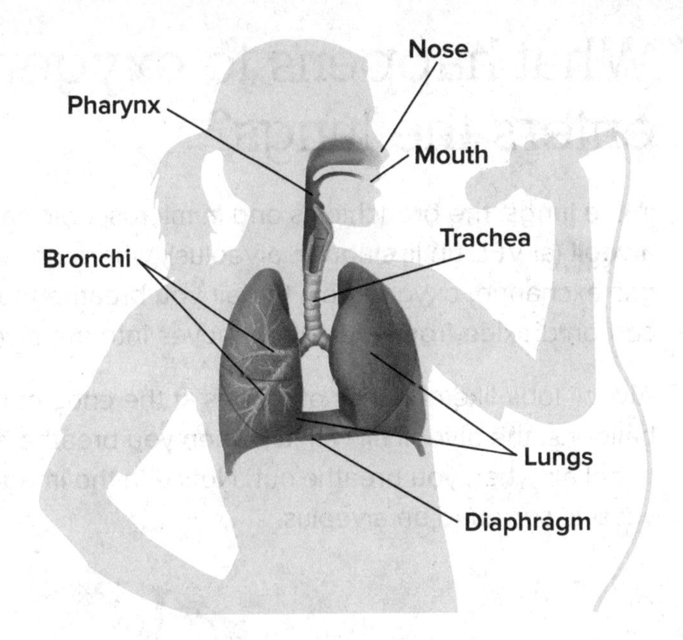

Oxygen enters the body when you inhale, or breathe in. Carbon dioxide leaves the body when you exhale. When you inhale, air enters the nostrils and passes through the pharynx. The **pharynx** is a tubelike passageway at the top of the throat that receives air, food, and liquids from the mouth or nose.

Because the pharynx is part of the throat, it is a part of both the digestive and respiratory systems. Food goes through the pharynx to the esophagus. Air travels through the pharynx to the **trachea,** a tube that is held open by C-shaped rings of cartilage. The trachea is also called the windpipe because it is a long, tubelike organ that connects the pharynx to the bronchi.

The **bronchi** are two narrower tubes that lead into the lungs. **Lungs** are the main organs of the respiratory system. Inside the lungs, the bronchi continue to branch into smaller and narrower tubes called bronchioles.

PHYSICAL SCIENCE Connection When high levels of carbon dioxide build up in your blood, the nervous system signals your body to breathe out. How does this happen?

The **diaphragm,** (DI uh fram) is a large muscle below the lungs that contracts and relaxes as air moves into and out of your lungs. For example, when your diaphragm contracts and moves down as you inhale, air rushes in to equalize air pressure. The diaphragm relaxes and moves up when you exhale. Air then rushes out to equalize the pressure. The movement of your diaphragm causes changes in the air pressure inside your chest. Breathing occurs because of these changes in air pressure.

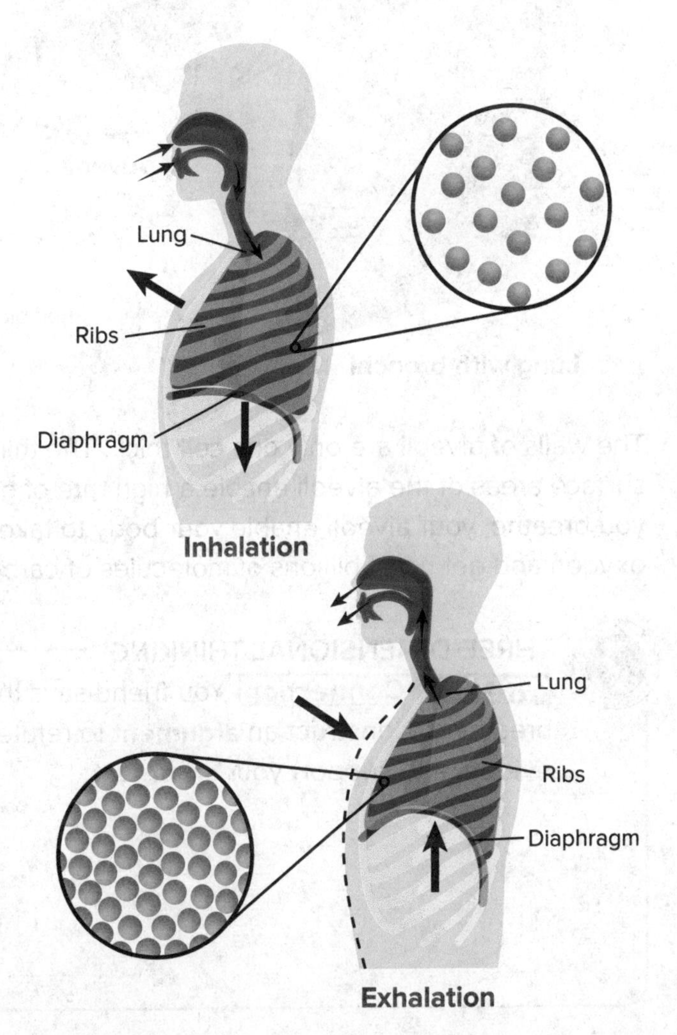

What happens to oxygen after it enters the lungs?

In the lungs, the bronchioles end in microscopic sacs, or pouches, called **alveoli** (al VEE uh li; singular, alveolus), where gas exchange occurs. During gas exchange, oxygen from the air you breathe moves into the blood, and carbon dioxide from your blood moves into the alveoli.

Alveoli look like bunches of grapes at the ends of the bronchioles. Like tiny balloons, the alveoli fill with air when you breathe in. They contract and expel air when you breathe out. Notice in the image below how blood vessels surround an alveolus.

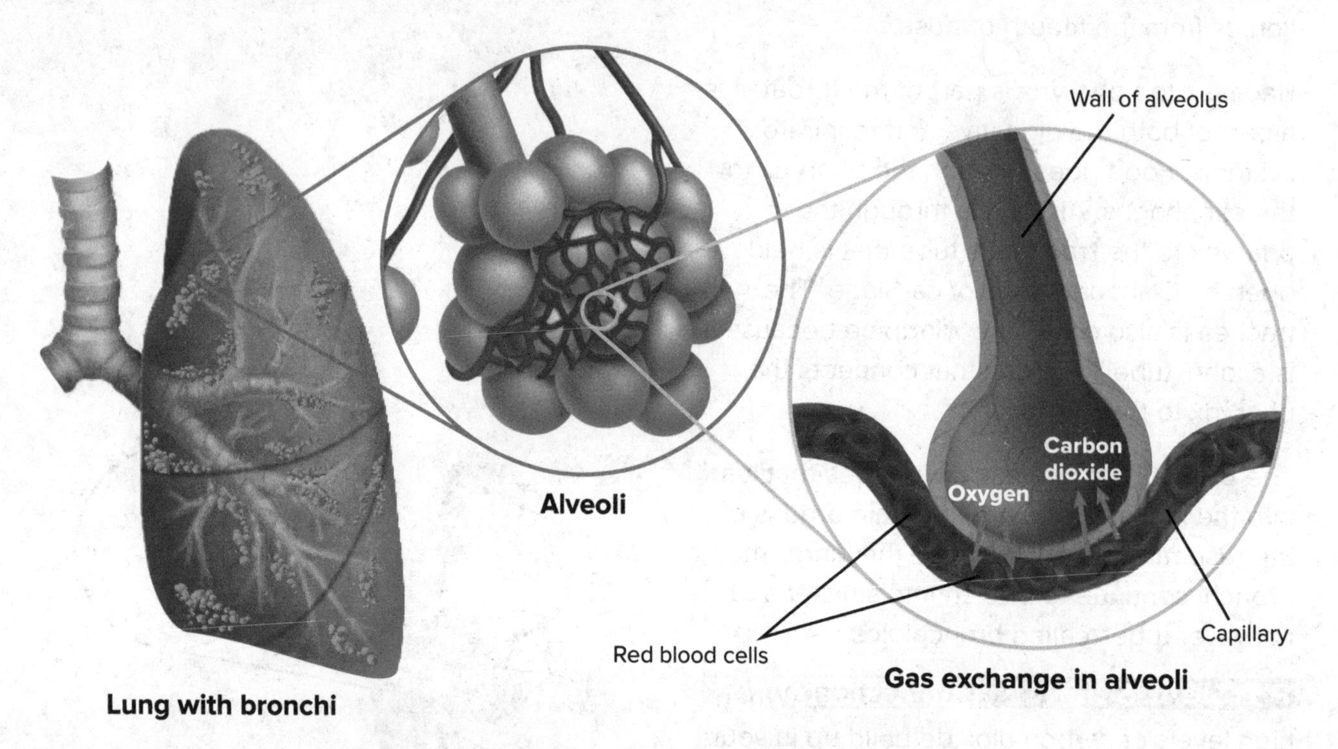

Lung with bronchi

Alveoli

Gas exchange in alveoli

The walls of alveoli are only one cell thick. The thin walls and the large surface areas of the alveoli enable a high rate of gas exchange. Every time you breathe, your alveoli enable your body to take in billions of molecules of oxygen and get rid of billions of molecules of carbon dioxide.

THREE-DIMENSIONAL THINKING

WRITING Connection You friend says that blood has no role in breathing. Construct an **argument** to refute his claim. Provide **evidence** to support your **argument.**

How is blood transported throughout the body?

Humans require oxygen to survive. All the cells in your body use oxygen to help process the energy in nutrients into energy that cells can use. You've learned that your lungs take in oxygen, and then oxygen moves into your blood. How does the oxygen in your blood get to the rest of your body?

INVESTIGATION

In a Heartbeat

1. Sit quietly for 1 minute.
2. Feel your pulse by placing the middle and index fingers of one hand on an artery in your neck or an artery in your wrist.
3. While sitting quietly, count the number of heartbeats you feel in 30 seconds. Multiply this number by two to calculate your pulse.
4. Record your data in your Science Notebook.
5. Jog in place for 1 minute.
6. Immediately repeat steps 3 and 4.
7. How did your pulse after exercising compare to your resting pulse?

8. Why do you think your pulse changed when you exercised?

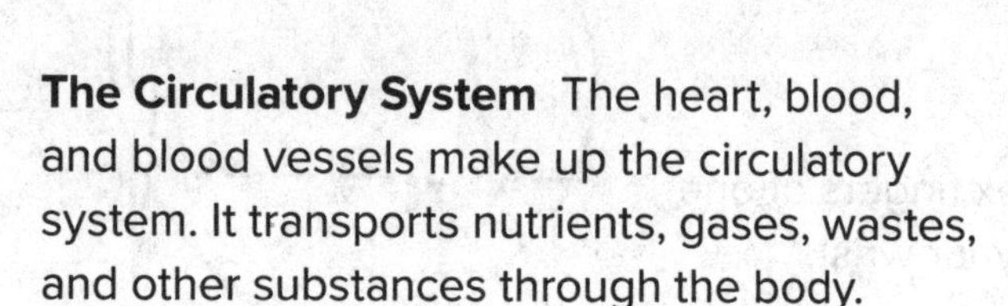

The Circulatory System The heart, blood, and blood vessels make up the circulatory system. It transports nutrients, gases, wastes, and other substances through the body.

Your heart is made up of muscle cells that constantly contract and relax. Contractions pump blood in your heart out of the heart and to the rest of your body. When your heart muscles relax, blood from the rest of your body enters the heart.

Look at the figure and notice that the heart has four chambers, two upper and two lower. Blood enters the upper two chambers of the heart, called the **atria** (AY tree uh; singular, atrium). Blood leaves through the lower two chambers of the heart, called the **ventricles.**

Blood travels through your body in tiny tubes called vessels. The three main types of blood vessels are arteries, veins, and capillaries. **Arteries** carry blood away from your heart. **Veins** transport blood that contains CO_2 back to your heart, except for the blood coming from the lungs, which is oxygen-rich. **Capillaries** are very tiny vessels that enable oxygen, CO_2, and nutrients to move between your circulatory system and your entire body.

COLLECT EVIDENCE

How do humans transport materials? Record your evidence (B) in the chart at the beginning of the lesson.

Read a Scientific Text

Body systems work hard to keep our bodies functioning, but they are not always able to function correctly. This can have many effects on a person's health. One result of poor function in the circulatory system is a heart attack.

CLOSE READING

Inspect

Read the text *Lower Heart Disease Risk.*

Find Evidence

Reread the passage. Underline the cause and effect of a heart attack.

Make Connections

Communicate Heart disease is the leading cause of death for Americans. With your partner, discuss how you can make choices to keep your heart healthy. Record your ideas below.

PRIMARY SOURCE

Lower Heart Disease Risk

Coronary heart disease—often simply called heart disease—is the main form of heart disease. It is a disorder of the blood vessels of the heart that can lead to heart attack. A heart attack happens when an artery becomes blocked, preventing oxygen and nutrients from getting to the heart. Heart disease is one of several cardiovascular diseases, which are diseases of the heart and blood vessel system. Other cardiovascular diseases include stroke, high blood pressure, angina (chest pain), and rheumatic heart disease.

[...] Heart disease is a lifelong condition—once you get it, you'll always have it. True, procedures such as bypass surgery and percutaneous coronary intervention can help blood and oxygen flow to the heart more easily. But the arteries remain damaged, which means you are more likely to have a heart attack. What's more, the condition of your blood vessels will steadily worsen unless you make changes in your daily habits. [...] That's why it is so vital to take action to prevent and control this disease.

Source: National Heart, Lung, and Blood Institute

STEM Careers

A Day in the Life of a Cardiovascular Technologist

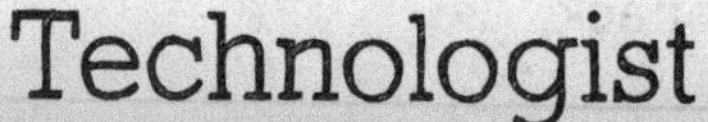

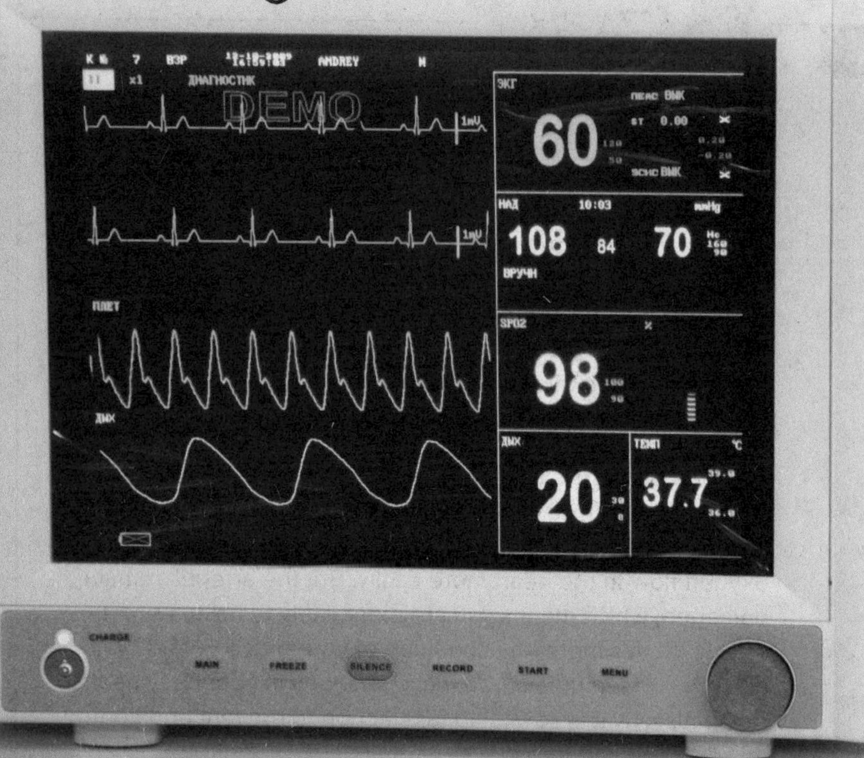

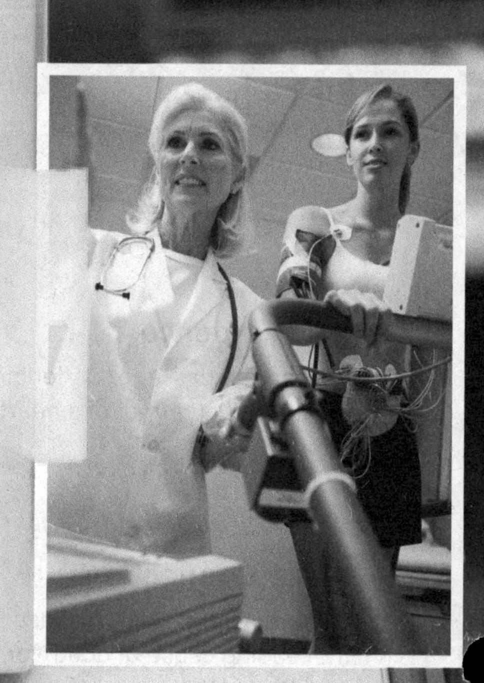

When you think about seeing a medical professional for your heart, you may think of a cardiologist, or even a heart surgeon. While these doctors will be the ones to conduct procedures, they have a lot of support staff, including cardiovascular technologists.

Cardiovascular technologists are trained to perform noninvasive diagnostic ultrasound, which provides digital video and images of anatomic data from inside the body. Cardiovascular technologists can help determine important diagnostic and treatment decisions for patients.

To enter the field of cardiovascular technology, you will need to complete a certification program at a hospital. You can also join this career by completing a two-year associate of science program.

It's Your Turn

Explain Would you like to become a cardiovascular technologist? Write a paragraph explaining why or why not, and then discuss your response with a partner.

LAB Modeling Blood Cells

Your body has different types of cells that perform various functions in the blood. Red blood cells carry oxygen to all the other cells in your body. White blood cells destroy viruses and bacteria that can attack the body and make you sick. These functions occur within your body everyday. How can you model the different functions of blood cells?

Safety    

Materials

toy cars and trucks
modeling clay
construction paper
glue stick
scissors
large sheet of paper
colored markers

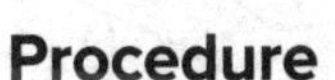

Procedure

1. Read and complete a lab safety form.
2. Cut out shapes from construction paper to represent the following organs: heart, lungs, stomach, and small intestine. Also cut out a shape to represent a body cell.
3. Draw an outline of a student on a large sheet of paper. Trace a fellow student if needed.
4. Place the organs in the appropriate body position on the outline. Choose a location away from the center of the body, such as an arm or a leg, to place the body cell.
5. Use the modeling clay to create molecules of oxygen, food, and waste materials (carbon dioxide and water). Place the oxygen molecules in the lungs. Place the food molecules in the stomach.
6. Your body gets energy when oxygen helps break down food molecules. Waste products are released during the breakdown of food molecules. Think about how a body cell gets energy.
7. Draw roads to connect the organs and the body cell so that the body cell can get the energy it needs. Select toy vehicles to represent red blood cells and white blood cells.

Procedure, continued

8. Draw a diagram of your model below.

9. How does oxygen reach body cells? Use the appropriate vehicle to model how red blood cells carry oxygen to a body cell. Add the path of the oxygen molecules to your diagram.

10. How do food molecules reach body cells? Use the appropriate vehicle to model how food molecules reach a body cell. Add the path of the food molecules to your diagram.

11. Where are waste materials produced? Use the appropriate vehicle to model how waste materials leave the body. Add the path to your diagram.

12. How is the body protected from viruses and bacteria? Use modeling clay to create viruses or bacteria and place them on your diagram. Use the appropriate vehicle to model how white blood cells destroy viruses and bacteria.

13. Follow your teacher's instructions for proper cleanup.

Analyze and Conclude

14. Explain why using police cars and red pickup trucks are appropriate models to represent white blood cells and red blood cells. Record your answer in your Science Notebook.

How do other animals take in oxygen and get rid of carbon dioxide?

All animals must take in oxygen and eliminate carbon dioxide to survive. Oxygen must enter the body so the cells and tissues are able to use it for life processes. Some animals have respiratory systems and breathe like humans do. However, other animals use various structures to perform gas exchange.

INVESTIGATION

Just Breathe

How do other animals breathe? With your group, research your assigned topic: diffusion, lungs, gills, or tracheal tubes. Once your research is complete, give a short presentation to teach your classmates what you have learned. Take notes on the other types of animal respiration as your classmates give their presentations.

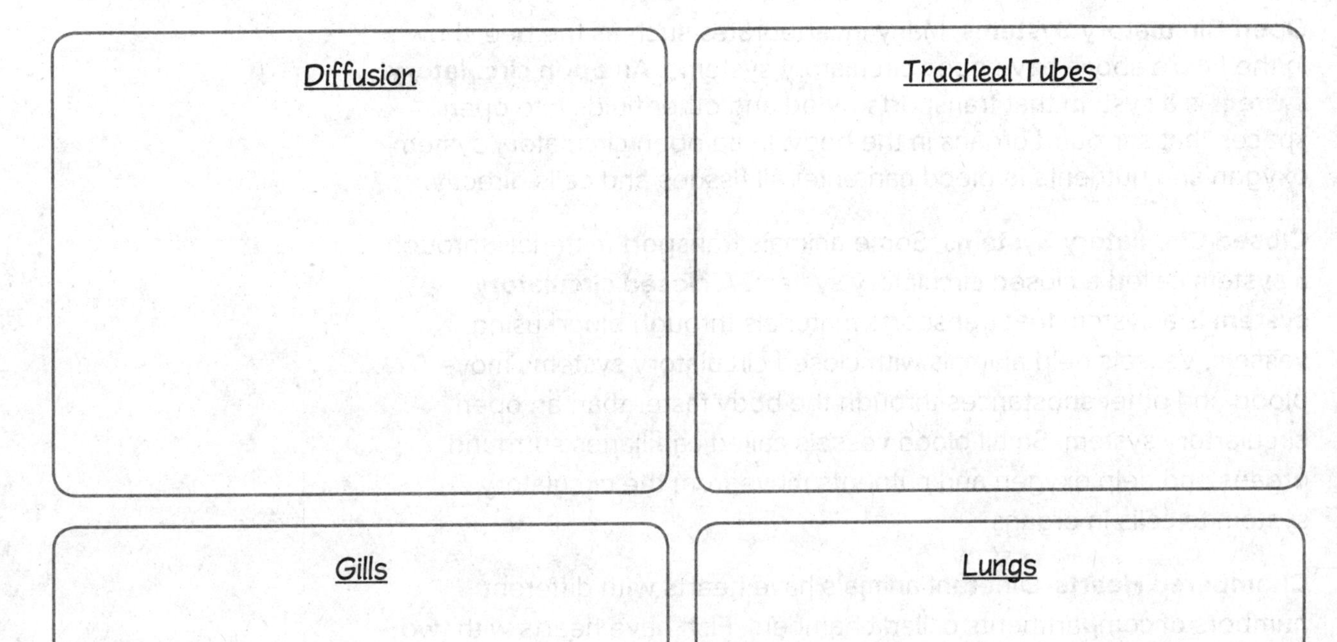

How do other animals transport blood throughout their bodies?

Different animals have different circulatory systems. The type of circulatory system used often determines how quickly blood moves through the animal.

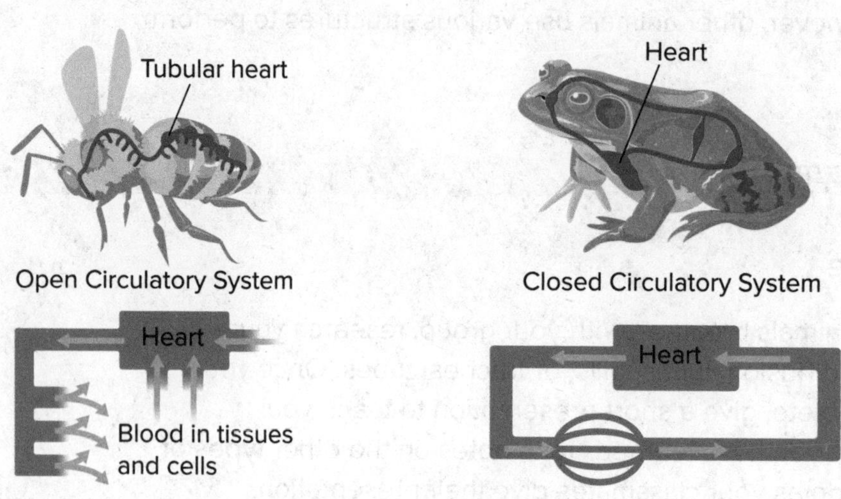

Open Circulatory Systems Many invertebrates such as the bee shown in the figure above have open circulatory systems. An **open circulatory system** is a system that transports blood and other fluids into open spaces that surround organs in the body. In an open circulatory system oxygen and nutrients in blood can enter all tissues and cells directly.

Closed Circulatory Systems Some animals transport materials through a system called a closed circulatory system. A **closed circulatory system** is a system that transports materials through blood using vessels. Vessels help animals with closed circulatory systems move blood and other substances through the body faster than an open circulartory system. Small blood vessels called capillaries surround organs and help oxygen and nutrients move from the circulatory system to cells in organs.

Chambered Hearts Different animals have hearts with different numbers of compartments called chambers. Fish have hearts with two chambers, whereas amphibian hearts consist of three chambers. Birds and mammals such as cats, dogs, and humans have hearts with four chambers. Almost all animals with three or four chambered hearts have lungs.

COLLECT EVIDENCE

How do other animals transport materials? Record your evidence (C) in the chart at the beginning of the lesson.

SCIENCE & SOCIETY

Very Special Blood Cells

Horseshoe crabs, living relatives of extinct trilobites, have been gathering on beaches for 350 million years. They usually become food for fish and birds. Yet someday your life might depend on horseshoe crabs—or at least on their blood. Unlike human blood, horseshoe crab blood contains only one type of blood cell. If bacteria enter the crab's bloodstream from an open wound, blood cells secrete a clotting factor. This secretion closes the wound, and the blood cells engulf the bacteria. When scientists saw that horseshoe crab blood turned to a gel in the presence of harmful bacteria, they realized its value. Today, medical professionals use an extract made from horseshoe crab blood to screen all intravenous medicines for bacteria. A quart of this special blood costs about $15,000!

The horseshoe crab blood can do even more. Another component of the blood can stop the human immunodeficiency virus (HIV) from replicating, or making copies of itself. Part of horseshoe crab blood can act as an antibiotic. Scientists also are using horseshoe crab blood in the development of a handheld instrument that helps to diagnose human illnesses. The instrument uses enzymes from the blood as illness detectors.

▲ **Technicians remove only a small portion of the crabs' blood. After this procedure, the crabs are returned to the ocean. Their blood cell levels return to normal in a couple of weeks.**

It's Your Turn

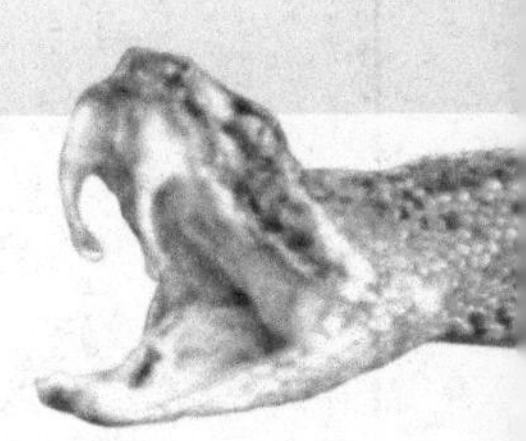

Report Medical professionals use certain types of snake venom to treat strokes. Research to find other unusual animal products that have medical uses, then write a paragraph about your findings to share with the class.

LESSON 4

Review

Summarize It!

1. **Identify** the functions of each human body system listed below.

Respiratory System

Circulatory System

Three-Dimensional Thinking

2. The arrow in the diagram below shows where blood enters the heart through the atrium after coming from the lungs. Which best describes the function of this blood entering the heart?

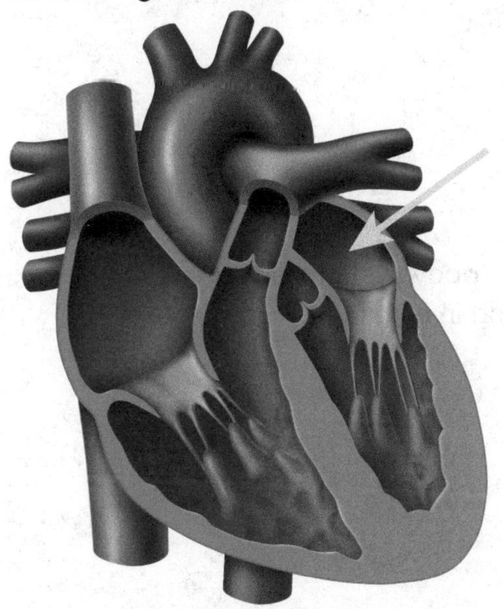

A The blood is carrying oxygen that it absorbed as it passed through the lungs.

B The blood is carrying carbon dioxide that it absorbed as it passed through the lungs.

C The blood is carrying nutrients that it absorbed as it passed through the small intestine.

D The blood is carrying capillaries that it absorbed as it passed through the stomach.

---

3. Which best explains the function of the alveoli in the respiratory system?

A The alveoli help to keep the lungs healthy by providing a way for all the cells in the lungs to obtain nutrients from the bloodstream.

B The alveoli help to keep the lungs inflated when you breathe out and make it possible to absorb oxygen when you breathe in.

C The alveoli provide a large surface area for absorbing oxygen from the air and releasing carbon dioxide wastes from the bloodstream.

D The alveoli provide a large surface area for absorbing oxygen from the air when you breathe in and also keep out harmful microorganisms.

Real-World Connection

4. **Describe** how the muscular system works with both the respiratory and circulatory systems when you are exercising.

5. **Refute** your friend's claim that plant body systems and animal body systems that transport materials have nothing in common.

Still have questions?
Go online to check your understanding about body systems that transport materials.

REVISIT

Do you still agree with the person you chose at the beginning of the lesson? Return to the Science Probe at the beginning of the lesson. Explain why you agree or disagree with that person now.

EXPLAIN THE PHENOMENON

Revisit your claim about how materials are transported in an organism. Review the evidence you collected. Explain how your evidence supports your claim.

KEEP PLANNING

STEM Module Project
Science Challenge

Now that you've learned about body systems that transport materials, go back to your Module Project to continue planning for your debate. You'll want to explain in your debate how these systems interact with other body systems in organisms, such as the glass frog.

PERFORMANCE EXPECTATION

LESSON 5 LAUNCH

When do we use our brains?

Three friends argued about the brain. They had different ideas about when they use their brains. This is what they said:

Abby: I think we use our brains only when we are conscious. When we are unconscious or asleep, another part of our nervous system takes over.

Laura: I think we use our brains when we are conscious, but only when we are doing something like thinking, speaking, eating, moving, and other intentional things. When we are unconscious or asleep, another part of our nervous system takes over.

Tony: I think we use our brains all the time, even when we are unconscious or asleep. The brain is always working.

Circle the person you agree with the most. Explain your thinking about the brain.

You will revisit your response to the Science Probe at the end of the lesson.

LESSON 5

Control and Information Processing

ENCOUNTER THE PHENOMENON | How does a surfer balance, move, and react to waves?

It can be hard to keep your balance when standing on one leg. Does shutting your eyes make this task easier or more difficult? Do you think surfers use their vision to help them stay on their surfboards?

1. Stand upright and lift your left leg, balancing yourself on your right leg.
2. Hold your left arm out so it is over your left knee.
3. Move your left leg backward and forward while maintaining your balance.
4. As you move your leg, swing your left arm back and forth at the same time. Have another student nearby to help you if you lose your balance.
5. Count how many times you are able to move your arm and your leg together before you lose your balance. Record this number below.
6. Repeat steps 2–4 with your eyes closed. Compare how many times you were able to swing your arm and your leg with your eyes open and with your eyes closed.
7. Was it easier to maintain your balance with your eyes open or closed? Explain your answer.

GO ONLINE Watch the video *Surf's Up!* to see this phenomenon in action.

EXPLAIN THE PHENOMENON

Did you see how it was easier to balance with your eyes open? Use your observations about the phenomenon to make a claim about how a surfer is able to balance, move, and react to a wave.

CLAIM

A surfer is able to balance, move, and react by....

COLLECT EVIDENCE as you work through the lesson.

Then return to these pages to record your evidence.

EVIDENCE

A. What evidence have you discovered to explain how the nervous system enables the surfer to balance, move, and react?

MORE EVIDENCE

B. What evidence have you discovered to explain how the senses enable the surfer to balance, move, and react?

When you are finished with the lesson, review your evidence. If necessary, based on the evidence, revise your claim.

REVISED CLAIM

A surfer is able to balance, move, and react by...

Finally, explain your reasoning for how and why your evidence supports your claim.

REASONING

The evidence I collected supports my claim because...

What controls the body's functions?

Have you ever started shivering because of the cold? Shivering is your body's way of keeping warm in response to cold temperatures. How are messages sent in the body in response to a stimulus, such as cold weather?

INVESTIGATION

Information Transportation

We can't actually see messages being sent through our bodies. In this activity, you will create a model for messages being sent through the body of a fish.

1. Form groups of five. Stand in a line with the four other students so that you can reach the hand of the person next to you. Do not move or talk. The student at the front of the line represents the nerve cells in a fish's head and eyes. The student at the end of the line represents the nerve cells in the fish's tail. Only the fish's tail—the last student in line—can make the fish swim.

2. Imagine that your fish sees a shrimp in the water. In order to eat the shrimp, the head must pass a message to the tail telling it to swim toward the shrimp. The student that represents nerve cells in the fish's head should send the message to the tail by squeezing the hand of the next student in line.

3. Continue passing the message between nerve cells (students). When the student representing nerve cells in the tail gets the message, this student should move forward. Observe how long it takes for the message to reach the tail and make the fish swim toward the shrimp.

4. Take one student out of the middle of the line, leaving a gap. Now imagine that the fish sees a shark that wants to eat it. What happens when the head sends a message to the tail to swim away from the shark? What happens to the fish?

5. Take two more students out of the middle of the line, making the line only two cells long—the fish's head and tail. Stand so that the head and tail are next to one another and hold hands. Now how long does it take to send a message to swim away from the shark? Why does the message move more quickly?

The Nervous System The part of an organism that gathers, processes, and responds to information is called the **nervous system.** The basic functioning units of the nervous system are called nerve cells, or **neurons.** The nervous system receives information from the senses, such as vision, hearing, smell, taste, and touch. Sense receptors respond to stimuli by sending signals that travel along neurons to the brain. The brain is the control center of your body. Your brain receives information, processes it, and sends out a response. The brain also stores some information as memories.

Your nervous system has two parts. The **central nervous system** (CNS) is made up of the brain and the spinal cord. The CNS receives, processes, stores, and transfers information. The **spinal cord** is a tubelike structure of neurons. The neurons extend to other areas of the body. This enables information to be sent out and received by the brain.

The **peripheral nervous system** (PNS) has sensory neurons and motor neurons that transmit information between the CNS and the rest of the body.

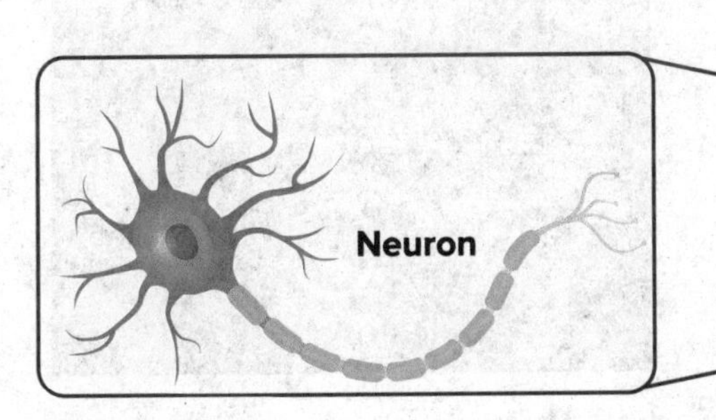

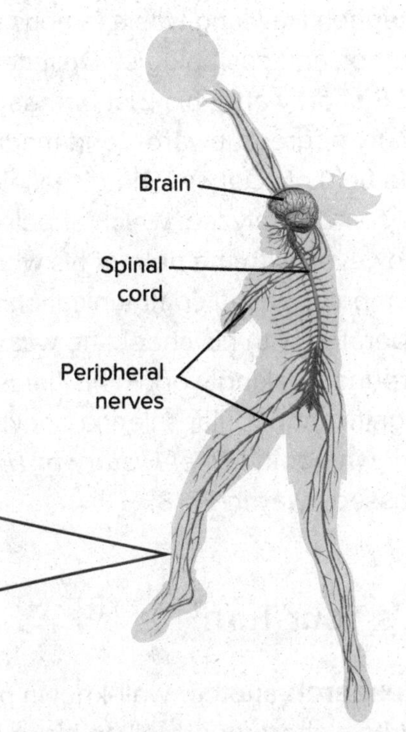

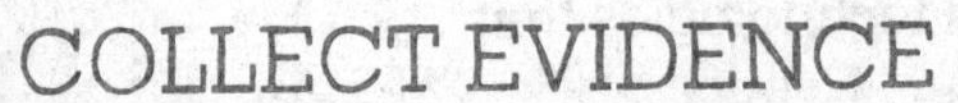

COLLECT EVIDENCE

How does the nervous system enable a surfer to balance, move and react? Record your evidence (A) in the chart at the beginning of the lesson.

A Closer Look: Amyotrophic Lateral Sclerosis

Amyotrophic lateral sclerosis, also known as ALS or Lou Gehrig's disease, affects cells in the brain and spinal cord, which in turn makes muscles weak and difficult to control. The disease progresses gradually and is different for every person affected.

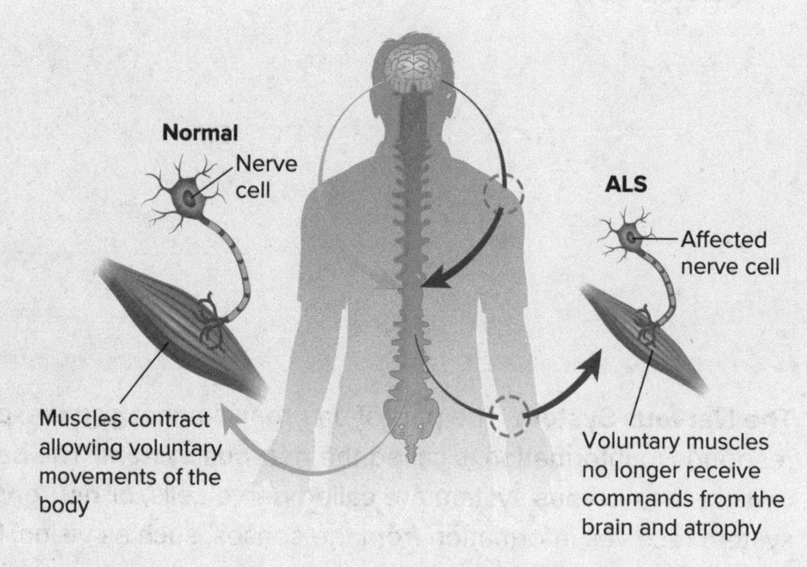

Symptoms of ALS can include muscle cramping, twitching, weakness, and difficulty speaking or swallowing. ALS is difficult to diagnose, and there are often many tests and procedures used during a diagnosis. There is currently no cure for ALS, but medicine and therapy can slow the progression of the disease and help ease symptoms. There are approximately 20,000 people in the United States living with ALS.

Stephen Hawking was a famous theoretical physicist, author, and cosmologist. Despite being diagnosed with ALS shortly after his 21st birthday, he went on to earn many degrees, awards, and medals for his contributions to the field of science. The disease caused him to gradually become paralyzed, which affected his ability to speak. However, with the help of his wheelchair and his computer-based communication system, which he operated with his cheek, he was able to produce groundbreaking work in his fields. He also wrote several significant popular science books including the well-known book *A Brief History of Time*. Stephen Hawking passed away in 2018.

It's Your Turn

Research another well-known person who has been affected by ALS. Write a short report telling his or her story, focusing on technologies that have helped him or her deal with living with the disease.

THREE-DIMENSIONAL THINKING

Explain how ALS, a disease of the nervous system, also affects other body systems.

How does your body receive information?

Your nervous system enables your body to receive, process, and react to information about your environment. Your nervous system is constantly responding to many different types of stimuli. However, your body has to receive a stimulus before it can respond to one. How does this happen? Brainstorm with a partner below.

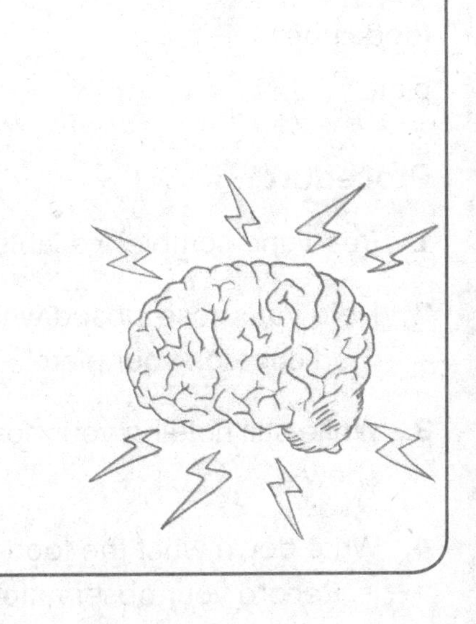

Want more information?

Go online to read more about body systems involved in control and information processing.

FOLDABLES

Go to the Foldables® library to make a Foldable® that will help you take notes while reading this lesson.

The Senses The **sensory system** is the part of your nervous system that detects or senses the environment. A human uses senses such as vision, hearing, smell, taste, and touch to detect his or her environment. There are also other senses, such as your sense of balance and sense of direction. All parts of the sensory system have special structures called **receptors** that detect stimuli. Each of the senses uses different receptors.

How does your body receive information through taste and smell?

Sometimes you might hold your nose when you don't want to taste something you're eating. Does this really work? Your senses all function because of nervous impulses that send information. Are your senses of taste and smell connected?

Taste with Your Nose

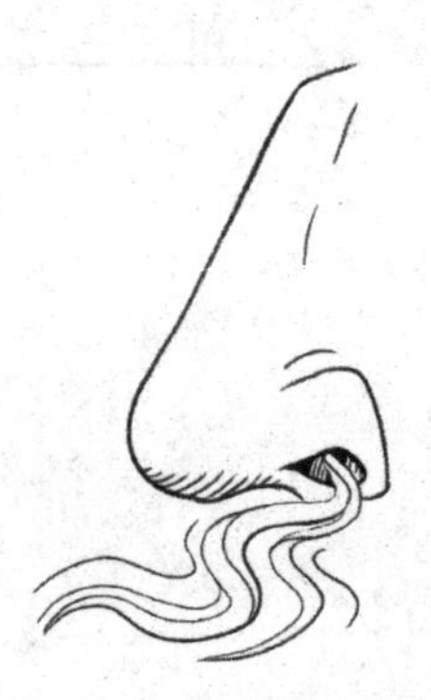

Materials

food cubes

plate

Procedure

1. Read and complete a lab safety form.
2. Hold your nose closed while your teacher walks around and places a food cube on your plate.
3. While still holding your nose, place the food cube in your mouth and chew it.
4. Write down what the food tasted like to you and what you think the food is. Record your observations in the Data and Observations section.
5. Repeat steps 2–4 with a different food cube. Record your observations.
6. Let go of your nose and chew another sample of both food cubes. Record your observations.
7. Follow your teacher's instructions for proper cleanup.

Data and Observations

Analyze and Conclude

8. What did the two food cubes taste like when you held your nose?

9. What did the two food cubes taste like when you were not holding your nose?

10. How do both senses together affect your ability to learn about your environment?

Taste and Smell Humans have hundreds of different receptors for detecting odors. Odors are molecules that are detected by chemical receptors, called chemoreceptors (kee moh rih SEP turz), in your nose. The sense of taste also relies on chemoreceptors. Chemoreceptors on your tongue detect chemicals in foods and drinks. Chemoreceptors on the tongue are called taste buds. The chemoreceptors in your nose and mouth work together to help you smell and taste foods. These receptors send messages to the brain. The brain then processes the information about the smell or taste and results in an immediate behavior (you spit out rotten food) or a memory (you remember the smell of rotten milk).

How does your body receive information through touch?

Your skin has millions of nerve endings. Which regions of the skin are the most sensitive to touch? How does the sense of touch help maintain homeostasis?

LAB Skin Sensitivity

Safety

Materials

clean tube sock or scarf

thin, washable markers of different colors

colored pencils

ruler

Procedure

1. Read and complete a lab safety form.
2. Blindfold your partner with a clean tube sock or scarf.
3. Lightly touch one of the areas of your partner's skin from the data table on the next page with a thin, washable marker. Leave a small, colored mark where you touched your partner's skin.

4. Using a different colored marker, have your partner try to touch the same point that was previously marked by your pen. Test all eight areas of the body listed on the table below (fingertip, back of hand, elbow, toes, top of foot, knee, back of neck, and nose). Use a ruler to determine the distance from the point you marked to the point your partner touched. Switch roles with your partner and repeat the process. Record your results and your partner's results in the data table below.

	Fingertip	Back of hand	Elbow	Toes	Top of foot	Knee	Back of neck	Nose
Person 1								
Person 2								

5. Plot the distance between the actual point of touch and the guessed point of touch and area of body data on the grid below. Plot the distance on the vertical axis and the area of the body on the horizontal axis. Label the axes and add a title to your plot. Use a different color of pencil for each set of data.

Procedure, continued

6. Follow your teacher's instructions for proper cleanup.

Analyze and Conclude

7. What part of your body has the least distance between the actual point and the guessed point? Which body part has the greatest distance?

8. Infer how the sensitivity of different areas of your skin help you maintain homeostasis.

Touch Like all the other senses, the sense of touch also uses special receptors that detect mechanical inputs in the environment. Receptors in your skin can detect temperature, pain, and pressure. Just like the other senses, touch receptors send messages to the brain where information is processed and memories are created. If you touch something hot and burn your hand, you'll remember how it felt.

SEP DCI CCC

THREE-DIMENSIONAL THINKING

Explain the **cause and effect** relationship involving touch and the ability to respond to an input or form a memory.

How does your body receive information through sound?

You already know that you need your ears in order to hear, but how exactly does hearing work? Complete the activity below to find out more about hearing.

INVESTIGATION

Hear Me Out

In this activity, you will research hearing with a partner. You will keep track of your sources in your Science Notebook as you research. Follow the citation format provided by your teacher. Make sure to use at least one print source and at least one digital source, and assess the credibility and accuracy of each source.

1. **READING Connection** What is sound? Research to find out. Quote or paraphrase your findings in order to avoid plagiarism.

2. **READING Connection** How does hearing work? Research to find out. Quote or paraphrase your findings in order to avoid plagiarism.

3. **MATH Connection** The formula for speed is:

$$\text{Speed (m/s)} = \frac{\text{Distance (m)}}{\text{Time (s)}}$$

The speed a sound wave travels depends on the type and temperature of the medium through which it is traveling. In air that is 68°F, the speed of sound is about 343 meters per second. Write the equation for the speed of sound using two variables.

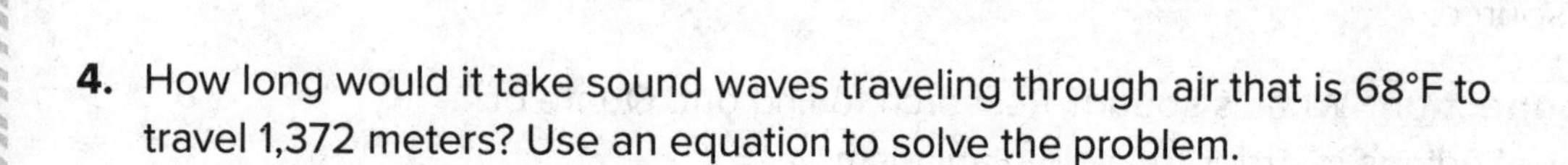

4. How long would it take sound waves traveling through air that is 68°F to travel 1,372 meters? Use an equation to solve the problem.

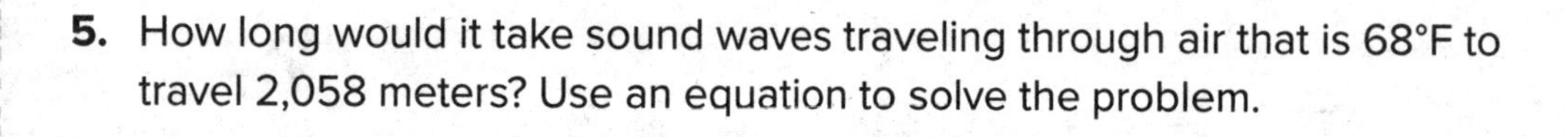

5. How long would it take sound waves traveling through air that is 68°F to travel 2,058 meters? Use an equation to solve the problem.

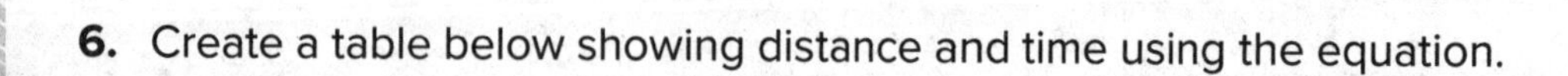

6. Create a table below showing distance and time using the equation.

7. Plot the distance and time data from the table on the previous page on the grid below. Plot the distance on the vertical axis and time on the the horizontal axis. Label the axes and add a title to your plot. Draw a line connecting the points.

8. Describe the relationship between the speed of sound, distance traveled, and time.

Hearing The vibration of matter creates mechanical sound waves that travel through air and other substances. Sound waves that enter the ear are detected by auditory (AW duh tor ee) receptors. As waves travel within the ear, they are amplified, or increased, and move hair cells. The hair cells send information about the sound waves to the brain. The brain processes information about the loudness and tone of the sound, and you hear. In addition to detecting sound waves, the inner ear also helps you maintain balance.

How does your body receive information through sight?

How does your sense of sight work? In the activity below, you will design a robot eye that engineers will use to allow robots to "see". You will only be allowed to use materials readily available at school or your home when building your eye. You must complete your assignment within the time frame given by your teacher.

ENGINEERING INVESTIGATION

Designing a Robot Eye

1. As a class, define the problem in this investigation.

2. List the criteria and constraints of the problem.

Criteria:	Constraints: Time: Materials:

3. **READING Connection** With your group, research the parts of the eye and describe their functions in the table on the next page.

Part of the eye	Function
Cornea	
Iris	
Pupil	
Lens	
Retina	
Optic nerve	
Ciliary muscle	

4. Decide as a group what materials you will use to build each part of your robot eye. Record your materials in the table below and provide reasoning to support your choices.

Part of the eye	Material	Reasoning
Cornea		
Iris		
Pupil		
Lens		
Retina		
Optic nerve		
Ciliary muscle		

5. With your group, create a scientific drawing of your eye. Include a front view and a cross section.

Front view:

Cross section:

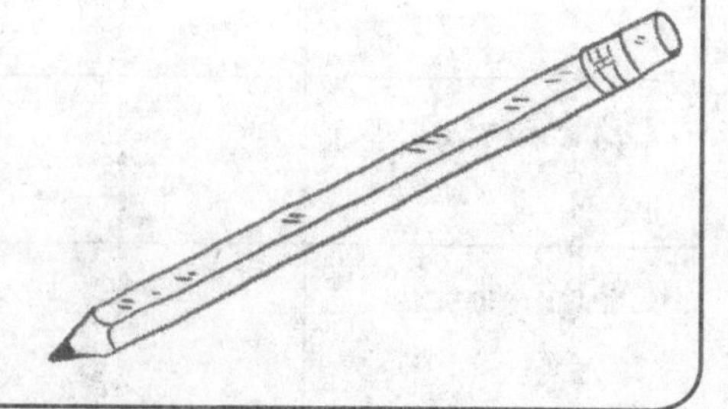

6. Create your interactive model.

7. Write an explanation for how your eye works.

8. Share your model and explanation with the class and observe the other group's models and explanations. Evaluate each group's design solutions. Do they meet the criteria and constraints of the problem?

Vision The visual system uses photoreceptors in the eye to detect electromagnetic signals—light—and create vision. Light enters the eye through the cornea (KOR nee uh), a thin membrane that protects the eye and changes the direction of light rays. The colored part of your eye is the iris (I rus). After light passes through the cornea, it goes through an opening formed by the iris called the pupil. The iris controls the amount of light that enters the eye by changing the size of the pupil. In bright light, the iris constricts, making the pupil smaller and letting in less light. In dim light, the iris relaxes, making the pupil larger and letting in more light.

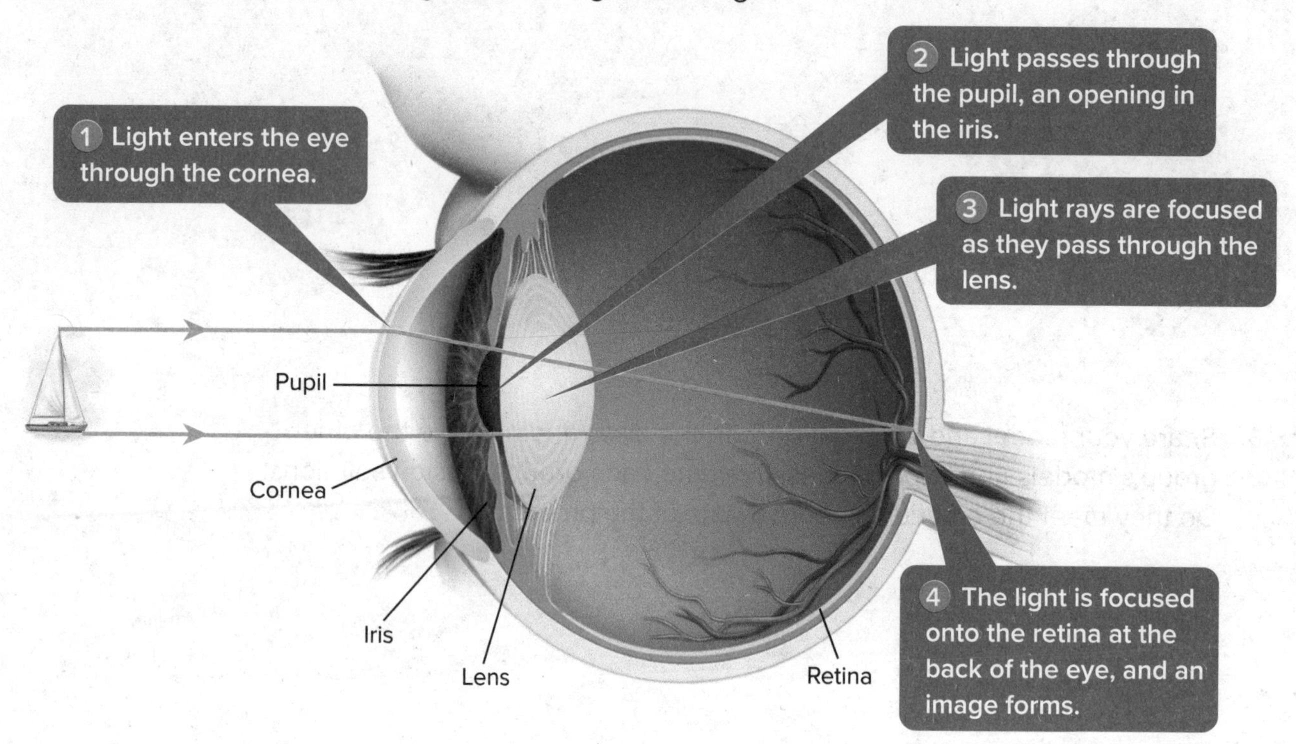

Light then travels through a clear structure called the lens. The lens works with the cornea and focuses light. The retina (RET nuh) is an area at the back of the eye that has two types of cells—rod cells and cone cells—with photoreceptors. The retina then sends information as electric signals through the optic nerve to the brain. The brain uses the information and creates a picture of what you are seeing.

THREE-DIMENSIONAL THINKING

How does the **model** your group created show the **structure and function** of the eye?

COLLECT EVIDENCE

How do the senses enable a surfer to balance, move, and react? Record your evidence (B) in the chart at the beginning of the lesson.

HOW IT WORKS

Night Vision Goggles

Would you like to be able to see in the dark?

You already know that vision begins with light entering your eyes. What happens when it is dark? If there is no light available to enter your eyes, how can any device make it possible to see?

Even when it is dark, there is light all around you. Although you can't see it, almost everything gives off, or emits, infrared light. Objects also reflect some infrared light, in the same way that they reflect visible light. Night vision goggles work by collecting that infrared light and converting it to visible light.

1 Infrared light enters the objective lens.

2 Infrared photons, or particles of light, enter the photocathode. This structure converts the pattern of infrared photons into a pattern of electrons.

3 The electrons speed up and are multiplied in the image intensifier tube.

4 The electrons strike a phosphor screen, a screen coated with phosphorescent material. The phosphor screen converts the electrons back into photons, forming an image that can be seen through the ocular lens.

It's Your Turn

READING Connection How can owls see more clearly in dim light than humans can? Research rods and cones in the eye and what they have to do with vision. Use what you discover to make a "How It Works" diagram about rods and cones in the eye.

How do plants respond to stimuli?

You just learned that humans and other animals use their senses, such as vision, touch, and hearing, to receive inputs from their environment. The senses detect stimuli, then trigger the nervous system to react. Are plants able to respond to stimuli too? Let's find out.

LAB Radish Research

Safety 

Materials

pot of young radish seedlings
toothpicks
light source

Procedure

1. Read and complete a lab safety form.
2. Choose a pot of young radish seedlings.
3. Place toothpicks parallel to a few of the seedlings in the pot in the direction of growth.
4. Place the pot near a light source, such as a gooseneck lamp or next to a window. The light source should be to one side of the pot, not directly above the plants. Illustrate the location of your pot and the light source to the right.
5. Check the position of the seedlings in relation to the toothpicks after 30 minutes. Record your observations in the Data and Observations section on the next page.
6. Observe the seedlings when you come to class the next day. Record your observations.
7. Follow your teacher's instructions for proper cleanup.

Data and Observations

Analyze and Conclude

8. What happened to the position of the seedlings after the first 30 minutes? What is your evidence of change?

9. What happened to the position of the seedlings after a day?

10. Why do you think the position of the seedlings changed?

Plant Responses to Stimuli Often a plant's response to stimuli might be so slow that it is hard to see it happen. The response might occur gradually over a period of hours or days. Light is a stimulus. A plant responds to light by growing toward it. This response occurs over several hours. In some cases, the response to a stimulus is quick, such as the Venus flytrap's response to touch. When stimulated by an insect's touch, the two sides of the trap snap shut immediately, trapping the insect inside. In addition to responding to light and touch, plants also respond to gravity.

LESSON 5

Review

Summarize It!

1. **Draw** a diagram showing how one type of signal in the environment is processed in the human body.

Three-Dimensional Thinking

Use the diagram below to answer question 2.

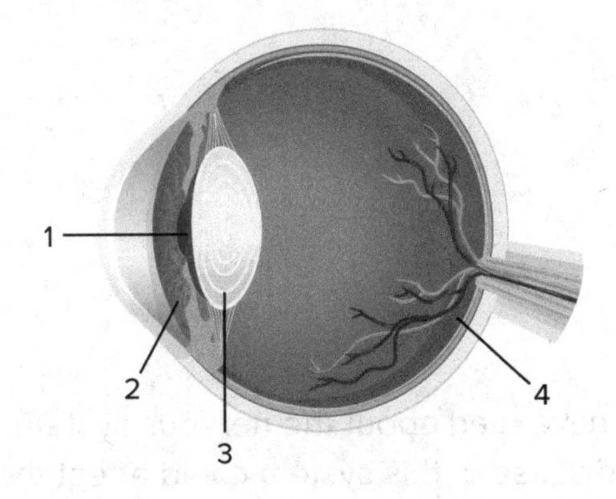

2. Which numbered structure focuses the light on the retina, where it is detected by photoreceptors?

A 1

B 2

C 3

D 4

---

3. A doctor sees a patient who has a loss of balance from an illness. The doctor thinks injury to the sense receptors for balance might be causing this effect. In which structure are they located?

A inner ear

B middle ear

C nasal cavity

D spinal cord

Real-World Connection

4. Describe how your senses help you form long term memories. Explain how this is beneficial to an organism.

5. Evaluate Using what you have read about the nervous system, evaluate whether a dysfunction or disease of this system could affect the functioning of the rest of the body. Explain your reasoning.

Still have questions?
Go online to check your understanding about body systems involved in control and information processing.

REVISIT

Do you still agree with the student you chose at the beginning of the lesson? Return to the Science Probe at the beginning of the lesson. Explain why you agree or disagree with that student now.

EXPLAIN
THE PHENOMENON

Revisit your claim about how a surfer can balance, move, and react to a wave. Review the evidence you collected. Explain how your evidence supports your claim.

PLAN AND PRESENT

STEM Module Project
Science Challenge

Now that you've learned about the nervous system and the senses, go back to your Module Project to finalize your debate. You'll want to explain in your debate how the nervous system and the senses work with other body systems in organisms, like the glass frog.

Body of Evidence

"Hey, Mr. Fernandez! We won our soccer game, thanks to my super strong muscles! I scored the winning goal!"

"That's great, Anna, but you know that you need more than your muscles to play soccer, right?"

Your team's task is to prepare to debate your classmate, who thinks that the body is made of independent subsystems that do not interact. You must provide evidence to support your argument that body systems in organisms, such as the soccer player and the glass frog, interact. Also provide information about how the senses impact the different body systems and enable the body to react and form memories.

Planning After Lesson 1

Fill in the path of organization from a cell to the organism.

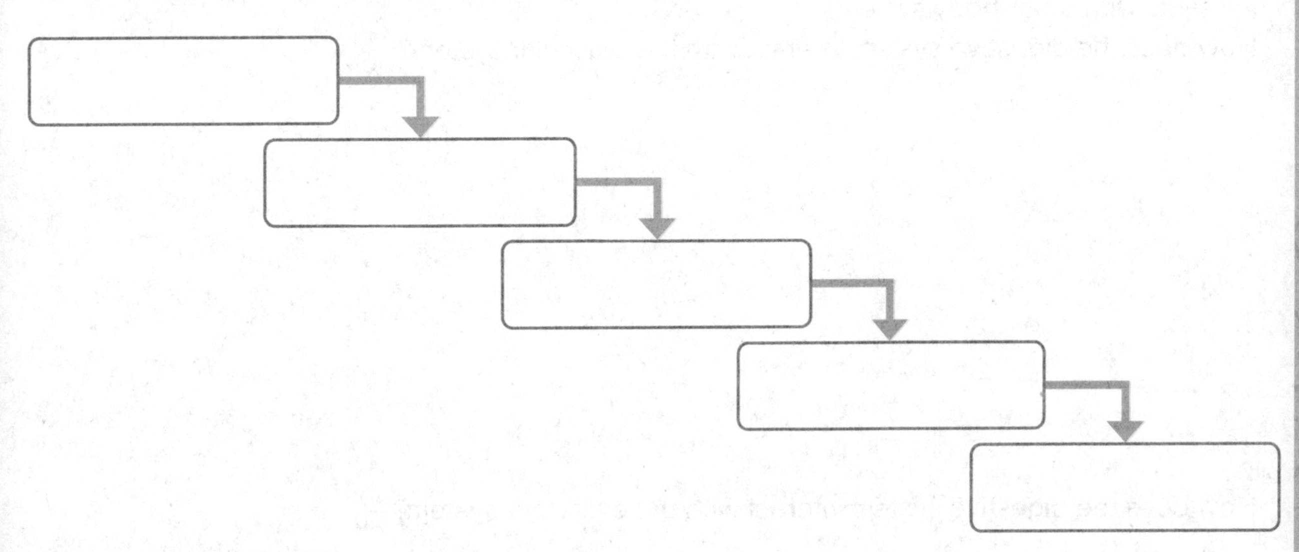

How is this organization similar in plant bodies? How is it different?

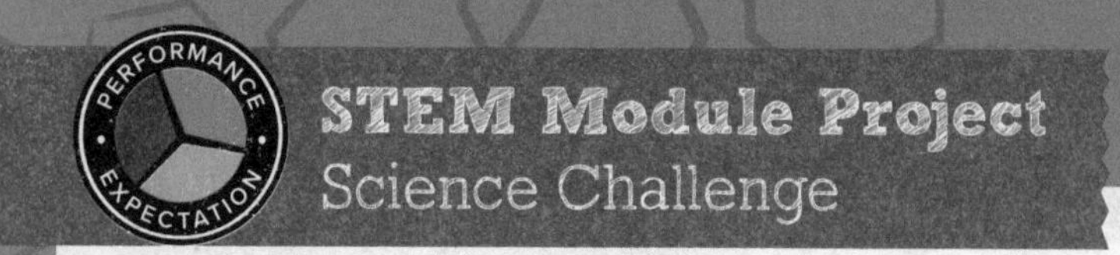

Planning After Lesson 2

Use these questions to help plan your argument that the muscular system interacts with other body systems:
How does the muscular system interact with the skeletal system?

How does the muscular system interact with the circulatory system?

Planning After Lesson 3

Use these questions to plan your argument that the digestive system interacts with other body systems:
How does the digestive system interact with the muscular system?

How does the digestive system interact with the excretory system?

Planning After Lesson 4

Use these questions to help plan your argument that body systems that transport materials interact with each other and interact with other body systems:
How does the respiratory system interact with the circulatory system?

How does the respiratory system interact with the muscular system?

Planning After Lesson 5

Explain the system that controls the body's functions. How will you argue in the debate that this system interacts with other body systems?

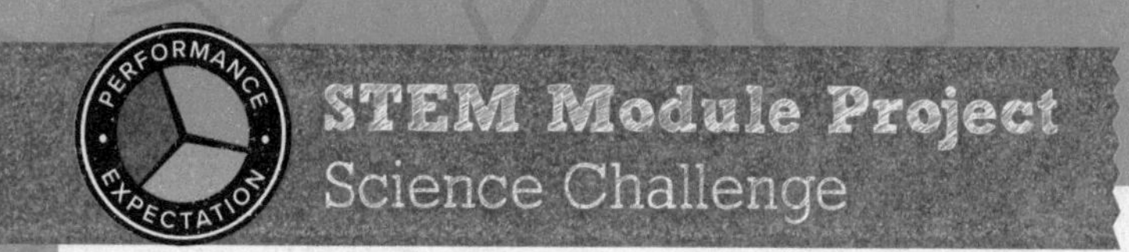

Write Your Script

Look back at the planning that you did after each lesson. Use that information to write a script for the argument that you will present during the debate.

Explain Your Argument

Now that you've planned your debate, complete the table below.

Debate
Claim What claim are you making about the phenomenon?
Obtaining Evidence and Information What information and evidence did you use to support your claim?
Evaluating Evidence and Information How did you ensure that the evidence and information used to support your claim is accurate and reliable?

STEM Module Project
Science Challenge

Present Your Argument

Analyze and evaluate your script before you head to the debate. How will your script lead your classmate to a better understanding of how body systems interact with one another?

How will your script lead your classmate to a better understanding of how the interaction of body systems enables organisms to perform life functions?

After listening to the other groups' presentations, return to your original script. What parts would you change or add after listening to other students?

Congratulations! You've completed the Science Challenge requirements!

Module Wrap-Up

REVISIT THE PHENOMENON

Using the concepts you learned throughout this module, explain how body systems in the glass frog work together in order to perform life functions.

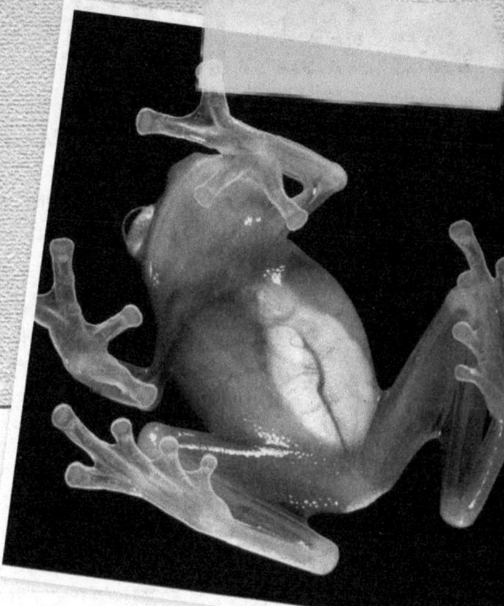

OPEN INQUIRY

If you had to ask one question about what you studied, what would it be?

Plan and conduct an investigation to answer this question.

Glossary

Multilingual Glossary

A science multilingual glossary is available on the science website. The glossary includes the following languages.

Arabic	Hmong	Tagalog
Bengali	Korean	Urdu
Chinese	Portuguese	Vietnamese
English	Russian	
Haitian Creole	Spanish	

Cómo usar el glosario en español:
1. Busca el término en inglés que desees encontrar.
2. El término en español, junto con la definición, se encuentran en la columna de la derecha.

Pronunciation Key

Use the following key to help you sound out words in the glossary.

a	back (BAK)	**Ew**	food (FEWD)
ay	day (DAY)	**yoo**	pure (PYOOR)
ah	father (FAH thur)	**yew**	few (FYEW)
ow	flower (FLOW ur)	**uh**	comma (CAH muh)
ar	car (CAR)	**u (+ con)**	rub (RUB)
E	less (LES)	**sh**	shelf (SHELF)
ee	leaf (LEEF)	**ch**	nature (NAY chur)
ih	trip (TRIHP)	**g**	gift (GIHFT)
i (i + com + e)	idea (i DEE uh)	**J**	gem (JEM)
oh	go (GOH)	**ing**	sing (SING)
aw	soft (SAWFT)	**zh**	vision (VIH zhun)
or	orbit (OR buht)	**k**	cake (KAYK)
oy	coin (COYN)	**s**	seed, cent (SEED)
oo	foot (FOOT)	**z**	zone, raise (ZOHN)

English — A — Español

alveolus/capillary

alveolo/capilar

alveolus (al VEE uh lus; plural, alveoli): microscopic sacs or pouches at the end of the bronchioles where gas exchange occurs.

alveolo (plural, alveolos): bolsas o sacos microscópicos en los extremos de los bronquiolos donde ocurre el intercambio de gas.

artery: a vessel that carries blood away from the heart.

arteria: vaso que lleva sangre fuera del corazón.

atria (AY tree uh; singular, atrium): the upper two chambers of the heart.

atrios (singular, atrio): las dos cámaras superiores del corazón.

B

bronchus (BRAHN kus; plural, bronchi): one of two narrow tubes that carry air into the lungs from the trachea.

bronquio (plural, bronquios): uno de los dos tubos delgados que llevan aire de la tráquea a los pulmones.

C

Calorie: the amount of energy it takes to raise the temperature of 1 kg of water by 1°C.

caloría: cantidad de energía necesaria para aumentar la temperatura de 1 kg de agua en 1°C.

capillary: a tiny blood vessel that delivers supplies to an individual cell and takes away waste materials.

capilar: vaso sanguíneo diminuto que entrega suministros a una célula individual y extrae los materiales de desecho.

cardiac (KAR dee ak) muscle: muscle found only in the heart.

cell differentiation (dihf uh ren shee AY shun): the process by which cells become different types of cells.

cell membrane: a flexible covering that protects the inside of a cell from the environment outside the cell.

cell theory: the theory that states that all living things are made of one or more cells, the cell is the smallest unit of life, and all new cells come from preexisting cells.

cell wall: a stiff structure outside the cell membrane that protects a cell from attack by viruses and other harmful organisms.

cell: the smallest unit of life.

cellular respiration: a series of chemical reactions that convert the energy in food molecules into a usable form of energy called ATP.

central nervous system (CNS): system made up of the brain and the spinal cord.

chemical digestion: a process in which chemical reactions break down pieces of food into small molecules.

chloroplast (KLOR uh plast): a membrane-bound organelle that uses light energy and makes food—a sugar called glucose—from water and carbon dioxide in a process known as photosynthesis.

closed circulatory system: a system that transports materials through blood using vessels.

cytoplasm: the liquid part of a cell inside the cell membrane; contains salts and other molecules.

músculo cardíaco: músculo que sólo se encuentra en el corazón.

diferenciación celular: proceso por el cual las células se convierten en diferentes tipos de células.

membrana celular: cubierta flexible que protege el interior de una célula del ambiente externo de la célula.

teoría celular: teoría que establece que todos los seres vivos están constituidos de una o más células (la célula es la unidad más pequeña de vida) y que las células nuevas provienen de células preexistentes.

pared celular: estructura rígida en el exterior de la membrana celular que protege la célula del ataque de virus y otros organismos dañinos.

célula: unidad más pequeña de vida.

respiración celular: serie de reacciones químicas que convierten la energía de las moléculas de alimento en una forma de energía utilizable llamada ATP.

sistema nervioso central (SNC): sistema constituido por el cerebro y la médula espinal.

digestión química: proceso por el cual las reacciones químicas descomponen partes del alimento en moléculas pequeñas.

cloroplasto: organelo limitado por una membrana que usa la energía lumínica para producir alimento –un azúcar llamado glucosa– del agua y del dióxido de carbono en un proceso llamado fotosíntesis.

sistema circulatorio cerrado: sistema que transporta materiales a través de la sangre usando vasos.

citoplasma: fluido en el interior de una célula que contiene sales y otras moléculas.

D

diaphragm (DI uh fram): a large muscle below the lungs that contracts and relaxes as air moves into and out of your lungs.

digestion: the mechanical and chemical breakdown of food into small particles and molecules that your body can absorb and use.

diafragma: músculo grande debajo de los pulmones que se contrae y relaja a medida que el aire entra y sale a los pulmones.

digestión: descomposición mecánica y química del alimento en partículas y moléculas pequeñas que el cuerpo absorbe y usa.

E

electron microscope: a microscope that uses a magnetic field to focus a beam of electrons through an object or onto an object's surface.

esophagus (ih SAH fuh gus): a muscular tube that connects the mouth to the stomach.

microscopio electrónico: microscopio que usa un campo magnético para enfocar un haz de electrones a través de un objeto o sobre la superficie de un objeto.

esófago: tubo muscular que conecta la boca al estómago.

excretory system: the system that collects and eliminates wastes from the body and regulates the level of fluid in the body.

sistema excretor: sistema que recolecta y elimina los desperdicios del cuerpo y regula el nivel de fluidos en el cuerpo.

exoskeleton: a thick, hard outer covering; protects and supports an animal's body.

exoesqueleto: cubierta externa, gruesa y dura; protege y soporta el cuerpo de un animal.

H

homeostasis (hoh mee oh STAY sus): an organism's ability to maintain steady internal conditions when outside conditions change.

homeostasis: capacidad de un organismo de mantener las condiciones internas estables cuando las condiciones externas cambian.

J

joint: where two or more bones meet.

articulación: donde dos o más huesos se unen.

L

ligament (LIH guh munt): the tissue that connects bones to other bones.

ligamento: tejido que conecta los huesos con otros huesos.

light microscope: a microscope that uses light and lenses to enlarge an image of an object.

microscopio de luz: microscopio que usa luz y lentes para aumentar la imagen de un objeto.

lungs: the main organs of the respiratory system.

pulmones: órganos principales del sistema respiratorio.

M

mechanical digestion: a process in which food is physically broken into smaller pieces.

digestión mecánica: proceso por el cual el alimento se descompone físicamente en pedazos más pequeños.

muscle: strong body tissue that can contract in an orderly way.

músculo: tejido corporal fuerte que se contrae de manera sistemática.

N

nervous system: the part of an organism that gathers, processes, and responds to information.

sistema nervioso: parte de un organismo que recoge, procesa y responde a la información.

neuron (NOO rahn): the basic functioning unit of the nervous system; a nerve cell.

neurona: unidad básica de funcionamiento del sistema nervioso; célula nerviosa.

nucleus: part of a eukaryotic cell that directs cell activity and contains genetic information stored in DNA.

núcleo: parte de la célula eucariótica que gobierna la actividad celular y contiene la información genética almacenada en el ADN.

O

open circulatory system: a system that transports blood and other fluids into open spaces that surround organs in the body.

sistema circulatorio abierto: sistema que transporta sangre y otros fluidos hacia espacios abiertos que rodean a los órganos en el cuerpo.

organ system: a group of organs that work together and perform a specific task.

sistema de órganos: grupo de órganos que trabajan juntos y realizar una función específica.

organ: a group of different tissues working together to perform a particular job.

órgano: grupo de diferentes tejidos que trabajan juntos para realizar una función específica.

organelle: membrane-surrounded component of a eukaryotic cell with a specialized function.

organelo: componente de una célula eucariótica rodeado de una membrana con una función especializada.

P

peripheral nervous system (PNS): system made of sensory and motor neurons that transmit information between the central nervous system (CNS) and the rest of the body.

sistema nervioso periférico (SNP): sistema formado por neuronas sensoriales y motoras que transmiten información entre el sistema nervioso central (SNC) y el resto del cuerpo.

peristalsis (per uh STAHL sus): waves of muscle contractions that move food through the digestive tract.

pharynx (FER ingks): a tubelike passageway at the top of the throat that receives air, food, and liquids from the mouth or nose.

phloem (FLOH em): a type of vascular tissue that carries dissolved sugars throughout a plant.

protein: a long chain of amino acid molecules; contains carbon, hydrogen, oxygen, nitrogen, and sometimes sulfur.

peristalsis: ondas de contracciones musculares que mueven el alimento por el tracto digestivo.

faringe: pasadizo parecido a un tubo en la parte superior de la garganta que recibe el aire, el alimento y los líquidos provenientes de la boca o de la nariz.

floema: tipo de tejido vascular que transporta azúcares disueltos por toda la planta.

proteína: larga cadena de aminoácidos; contiene carbono, hidrógeno, oxígeno, nitrógeno y, algunas veces, sulfuro.

R

receptor: special structures in all parts of the sensory system that detect stimuli.

receptor: estructuras especiales en todas partes del sistema sensorial que detectan los estímulos.

S

sensory system: the part of your nervous system that detects or senses the environment.

skeletal system: body system that contains bones as well as other structures that connect and protect the bones and that support other functions in the body.

smooth muscle: involuntary muscle named for its smooth appearance.

spinal cord: a tubelike structure of neurons that sends signals to and from the brain.

stoma (STOH muh): a small opening in the epidermis, or surface layer, of a leaf.

sistema sensorial: parte del sistema nervioso que detecta o siente el medioambiente.

sistema esquelético: sistema corporal que comprende los huesos al igual que otras estructuras que conectan y protegen los huesos y que apoyan otras funciones en el cuerpo.

músculo liso: músculo involuntario llamado así por su apariencia lisa.

médula espinal: estructura de neuronas en forma de tubo que envía señales hacia y del cerebro.

estoma: abertura pequeña en la epidermis, o capa superficial, de una hoja.

T

tissue: a group of similar types of cells that work together to carry out specific tasks.

trachea (TRAY kee uh): a tube that is held open by C-shaped rings of cartilage; connects the larynx and the bronchi.

tejido: grupo de tipos similares de células que trabajan juntas para llevar a cabo diferentes funciones.

tráquea: tubo que los anillos en forma de C del cartílago mantienen abierto; este conecta la laringe y los bronquios.

U

unicellular: a living thing that is made up of only one cell.

unicelular: ser vivo formado por una sola célula.

V

vascular tissue: specialized plant tissue composed of tubelike cells that transports water and nutrients in some plants.

vein: a vessel that carries blood toward the heart.

ventricles (VEN trih kulz): the lower two chambers of the heart.

tejido vascular: tejido especializado de la planta compuesto de células tubulares que transportan agua y nutrientes en algunas plantas.

vena: vaso que lleva sangre hacia el corazón.

ventrículos: las dos cámaras inferiores del corazón.

villus (VIH luhs): a fingerlike projection, many of which cover the folds of the small intestine.

vellosidad: proyección parecida a un dedo, muchas de las cuales cubren los pliegues del intestino delgado.

X

xylem (ZI lum): a type of vascular tissue that carries water and dissolved nutrients from the roots to the stem and the leaves.

xilema: tipo de tejido vascular que transporta agua y nutrientes disueltos desde las raíces hacia el tallo y las hojas.

Index

Italic numbers = illustration/photo
Bold numbers = vocabulary term
lab = indicates entry is used in a lab
inv = indicates entry is used in an investigation
smp = indicates entry is used in a STEM Module Project
enc = indicates entry is used in an Encounter the Phenomenon
sc = indicates entry is used in a STEM Career